**Enabling 5G Communication Systems to Support
Vertical Industries**

Enabling 5G Communication Systems to Support Vertical Industries

Edited by

Muhammad Ali Imran
University of Glasgow
UK

Yusuf Abdulrahman Sambo
University of Glasgow
UK

Qammer H. Abbasi
University of Glasgow
UK

Registered Offices
John Wiley & Sons, Inc., 111 River Street, Hoboken, NJ 07030, USA
John Wiley & Sons Ltd, The Atrium, Southern Gate, Chichester, West Sussex, PO19 8SQ, UK

Editorial Office
The Atrium, Southern Gate, Chichester, West Sussex, PO19 8SQ, UK

For details of our global editorial offices, customer services, and more information about Wiley products visit us at www.wiley.com.

Wiley also publishes its books in a variety of electronic formats and by print-on-demand. Some content that appears in standard print versions of this book may not be available in other formats.

Library of Congress Cataloging-in-Publication data applied for

Hardback ISBN: 9781119515531
A catalogue record for this book is available from the British Library.

Cover Design: Wiley
Cover Image: © Photographer is my life / Getty Images

Set in 10/12pt WarnockPro by SPi Global, Chennai, India
Printed and bound in Singapore by Markono Print Media Pte Ltd

10 9 8 7 6 5 4 3 2 1

Contents

About the Editors

Muhammad Ali Imran is the Vice Dean Glasgow College UESTC and Professor of Communication Systems in the School of Engineering at the University of Glasgow. He leads the Communications, Sensing and Imaging research group. He was awarded his MSc (Distinction) and PhD degrees from Imperial College London, UK, in 2002 and 2007, respectively. He is an Affiliate Professor at the University of Oklahoma, USA and a visiting Professor at 5G Innovation Centre, University of Surrey, UK. He has over 18 years of combined academic and industry experience, working primarily in the research areas of cellular communication systems. He has been awarded 15 patents, has authored/co-authored over 300 journal and conference publications, and has been principal/co-principal investigator on over £6 million in sponsored research grants and contracts. He has supervised 30+ successful PhD graduates. He has an award of excellence in recognition of his academic achievements, conferred by the President of Pakistan. He was also awarded IEEE Comsoc's Fred Ellersick Award 2014, FEPS Learning and Teaching Award 2014, and Sentinel of Science Award 2016. He was twice nominated for Tony Jean's Inspirational Teaching Award. He is a shortlisted finalist for The Wharton-QS Stars Awards 2014, QS Stars Reimagine Education Award 2016 for innovative teaching and VC's learning and teaching award in University of Surrey. He is the co-editor of two books: *Access, Fronthaul and Backhaul Networks for 5G and Beyond*, published by IET, ISBN 9781785612138, and *Energy Management in Wireless Cellular and Ad-hoc Networks*, published by Springer, ISBN 9783319275666. He is a senior member of IEEE, Fellow of IET and a Senior Fellow of Higher Education Academy (SFHEA), UK.

Yusuf Abdulrahman Sambo received the MSc degree (Distinction) in Mobile and Satellite Communications, in 2011, and the PhD degree in electronic engineering in 2016, from the Institute for Communication Systems (ICS, formally known as CCSR) of the University of Surrey. He is currently a Postdoctoral Research Associate in the mobility, massive Internet Communications, Sensing and Imaging (CSI) research group at the University of Glasgow. Prior to joining the University of Glasgow, he was a Lecturer in Telecommunications Engineering at Baze University, Abuja from June 2016 to September 2017. His main research interests include self-organized networks, radio resource management, EM exposure reduction, energy efficiency and 5G testbed implementation. He is a member of IEEE and the IET. He has served as technical program committee member of several IEEE conferences and as reviewer for several IEEE and other top journals. He has also contributed in organizing IEEE conferences and workshops.

Qammer H. Abbasi received his BSc and MSc degrees in electronics and telecommunication engineering from University of Engineering and Technology (UET), Lahore, Pakistan (with Distinction). He received his Ph.D. degree in Electronic and Electrical engineering from Queen Mary University of London (QMUL), UK, in January, 2012. Until June 2012, he was Postdoctoral Research Assistant in Antenna and Electromagnetics group, QMUL, UK. From 2012 to 2013, he was international young scientist under National Science Foundation China (NSFC), and Assistant Professor in University of Engineering and Technology (UET), KSK, Lahore. From August, 2013 to April 2017 he was with the Centre for Remote Healthcare Technology and Wireless Research Group, Department of Electrical and Computer Engineering, Texas A&M University (TAMUQ), initially as an Assistant Research Scientist and later was promoted to an Associate Research Scientist and Visiting Lecturer. Currently Dr Abbasi is a Lecturer (Assistant Professor) in University of Glasgow in the School of Engineering in addition to Visiting Lecturer (Assistant Professor) with Queen Mary University of London (QMUL). Dr Abbasi has grant portfolio of around £3.5 million, contributed to a patent and more than 180 leading international technical journal and peer-reviewed conference papers, in addition to five books, and received several recognitions for his research. His research interests include nano communication, the Internet of Things, 5G and its applications to connected health, RF design and radio propagation, applications of millimetre and terahertz communication in healthcare and agri-tech, wearable and flexible sensors, compact antenna design, antenna interaction with human body, implants, body-centric wireless communication issues, wireless body sensor networks, non-invasive healthcare solutions and physical layer security for wearable/implant communication. Dr Abbasi is an IEEE senior member and was Chair of IEEE young professional affinity group. He is an Associate Editor for *IEEE Access* journal, *IEEE Journal of Electromagnetics, RF and Microwaves in Medicine and Biology, Advanced Electromagnetics Journal* and acted as a guest editor for numerous special issues in top-notch journals including Elsevier's *Nano Communication Networks*. He is member of the IEEE 1906.1.1 standard committee on nano communication and is also a member of IET and committee member for the IET Antenna & Propagation and healthcare network. Dr Abbasi has been a member of the technical program committees of several IEEE flagship conferences and technical reviewer for several IEEE and top-notch journals. He contributed in organizing several IEEE conferences, workshop and special sessions in addition to a European School of Antennas course.

List of Contributors

Wasim Ahmad
School of Engineering
University of Glasgow
Glasgow

Nauman Aslam
Department of Computer Science and
Digital Technologies
Northumbria University
Newcastle

Konstantinos Antonakoglou
Department of Informatics
Kings College London

Islam Safak Bayram
Qatar Environment and Energy Research
Institute (QEERI)
Hamad Bin Khalifa University
Qatar

Mehrdad Dianati
Warwick Manufacturing Group (WMG)
University of Warwick
Coventry

Fabrizio Granelli
Department of Engineering and
Information Science
University of Trento
Italy

Chong Han
pureLiFi
Edinburgh

Sajjad Hussain
School of Engineering
University of Glasgow
Glasgow

Ali Imran
Artificial Intelligence for Networks
(AI4Networks) Lab
School of Electrical and Computer
Engineering
University of Oklahoma
Tulsa, USA

Matteo Innocenti
Azienda Ospedaliero-Universitaria
Careggi
Firenze
Italy

Muhammad Ismail
Electrical and Computer Engineering
Department
Texas A&M University at Qatar
Qatar

Kostas Katsaros
Digital Catapult
London

Stamos Katsigiannis
School of Engineering and Computing
University of the West of Scotland
Paisley

Paulo Valente Klaine
School of Engineering
University of Glasgow
Glasgow

Hoa Le-minh
Department of Mathematics, Physics and
Electrical Engineering
Northumbria University
Newcastle

Yi Lu
Warwick Manufacturing Group (WMG)
University of Warwick
Coventry

Maliheh Mahlouji
Department of Informatics
Kings College London

Toktam Mahmoodi
Centre for Telecommunications Research
Department of Informatics
Kings College London

Carsten Maple
Warwick Manufacturing Group (WMG)
University of Warwick
Coventry

Alex Mouzakitis
Department of Electrical, Electronics and
Software Engineering Research
Jaguar Land Rover
Coventry

João Pedro Battistella Nadas
School of Engineering
University of Glasgow
Glasgow

Huan Nguyen
Department of Design Engineering
and Mathematics
Middlesex University London

Rafaela de Paula Parisotto
Department of Electrical and
Electronics Engineering
Federal University of Santa Catarina
Brazil

Khalid Qaraqe
Electrical and Computer Engineering
Department
Texas A&M University at Qatar
Qatar

Haneya Naeem Qureshi
Artificial Intelligence for Networks
(AI4Networks) Lab
School of Electrical and Computer
Engineering
University of Oklahoma
Tulsa, USA

Naeem Ramzan
School of Engineering and Computing
University of the West of Scotland
Paisley

Mohsin Raza
Department of Design Engineering
and Mathematics
Middlesex University London

Usman Raza
Toshiba Research Europe Limited
Cambridge

Erchin Serpedin
Electrical and Computer Engineering
Department
Texas A&M University
College Station
Texas, USA

Luis Sequeira
Department of Informatics
Kings College London

David Soldani
Huawei
Australia

Richard D. Souza
Department of Electrical and
Electronics Engineering
Federal University of Santa Catarina
Brazil

Andy Sutton
British Telecom
Warrington
United Kingdom

Muhammad Usman
Department of Engineering and
Information Science
University of Trento
Italy

Preface

Recent technological advances have resulted in the creation of vertical industries that provide niche services that were hitherto non-existent only a decade ago. These industries are highly dependent on the availability and reliable exchange of data between multiple locations. As such, there is a consensus among analysts and opinion leaders that data will be the key driver in the fourth industrial revolution. Given the scale of devices to be connected in each vertical industry and the vast amount of data to be exchanged between multiple points, it is obvious that wireless communication is the leading enabler of the services envisioned by these industries. However, the current wireless communication systems have several technical limitations that would inhibit the actualization of the targets of these vertical industries. Hence, it is paramount to redesign wireless communication systems to take into account the holistic operational requirements of the vertical industries to enable them to reach the new productivity domains.

Mobile communication devices have rapidly evolved to become ubiquitous in our everyday lives. Although they were generally used for voice communication in the early days of the mobile communication system, steady advancement in research has paved the way for a significant increase in Spectral Efficiency (SE), which resulted in a paradigm shift towards wireless data communication. This transition has led to an increase in data rate from 50 kbps in 2G systems to 1 Gbps in the current 4G systems. Unfortunately, the popularity of wireless communications and the increasing demand for high data rates comes with the challenge of a dramatic increase in the energy consumption of these networks. Consequently, researchers had to focus on designing energy-efficient networks that would increase the amount of bits transmitted per joule of energy or reduce the energy required to transmit a bit of data.

Unlike existing generations of wireless communication systems that have simply been upgrades mostly in terms of SE and Energy Efficiency (EE), the fifth-generation mobile communication network (5G) seeks to bring about a revolution in the way mobile systems are perceived. It is envisioned that augmented reality, virtual reality, tele-robotics, remote surgery and haptic communication will be the revolutionary applications of 5G. Accordingly, 5G promises to stretch the limits of the Key Performance Indicators (KPIs) of current systems by taking into account several criteria such as latency, resilience, connection density and coverage area, alongside the traditional SE and EE criteria. 5G has design targets of sub-millisecond end-to-end latency, 100-fold increase in typical user data rates, 100 times increase in connection density and 10 times increase in EE, compared to current systems. This makes 5G the prime candidate to support a wide range of vertical industries.

Given that each vertical industry has its specific requirements for optimal service delivery, the 5G communication system will have to provide tailor-made solutions for each industry against the "one size fits all anywhere, anytime" approach of current systems. In order to achieve this, novel techniques such as Software Defined Networking (SDN), Network Function Virtualization (NFV), new waveforms, carrier aggregation and network slicing would form the basis of 5G.

This book evaluates advances in the current state-of-the-art and provides readers with insights on how 5G can seamlessly support vertical industries. It explores the recent advances in theory and practice of 5G and beyond communication systems that provide support for specific industrial sectors such as smart transportation, connected industries, e-healthcare, smart grid, media and entertainment and disaster management. Furthermore, this book highlights how 5G and beyond communication systems can accommodate the unique frameworks and Quality of Service (QoS) requirements of vertical industries for efficient and cost-effective service delivery. Each chapter in the book is designed to focus on how 5G would enable a specific vertical industry.

Chapter 1 presents 5G network architecture, design and service optimization. The author introduces the latest Third Generation Partnership Project (3GPP) release of 5G (Release 15), which contains specifications of a new radio interface that connects to an enhanced evolved packet core, referred to as EPC+. The author also sheds light on the expected specifications of 3GPP Release 16, which will fully exploit 5G for vertical industries. Furthermore, the author describes 5G use cases families as identified by the Next Generation Mobile Networks (NGMN), which are broadband access in dense areas, broadband accessibility everywhere, higher user mobility, massive Internet of Things (IoT), extreme real-time communications, lifeline communications, ultra-reliable communications and broadcast-like services. These use cases enable support for various vertical industries. 5G network architecture is also introduced, with the author describing the new functional blocks, interfaces, control and user plane splitting, radio access network, as well as the integration of NFV and SDN to form a virtualized architecture. The chapter also covers the process of deploying a 5G network overlaid on current networks. The author concludes the chapter by proclaiming, "There is not an industry or business sector that will not be impacted by the introduction of 5G in support of an increasingly connected and automated digital workplace."

In Chapter 2, the authors provide a comprehensive description of technology improvements, industrial processes and control requirements for the fourth industrial revolution. The authors make a case for and provide insights into how wireless sensor networks of the future can benefit from novel 5G technologies to improve efficiency in industrial processes. The chapter compares the typical QoS requirements of industrial applications based on battery life, security and update frequency, and then presents a detailed overview of the Industrial Wireless Sensor Network (IWSN) architecture, including node specifications, network topologies and channel access strategies. After highlighting the importance of Ultra-Reliable Low Latency Communication (URLLC) for IWSNs, the chapter proposes a hybrid multi-channel scheme for performance and throughput enhancement of IWSNs whereby multiple frequency channels are used to implement URLLC in IWSNs. The scheme utilizes a novel priority-based scheduler for retransmission of failed packets on a special frequency channel, and frequency polling to mitigate collisions. This approach brings about a reduction in frame error rate, and improvement in reliability.

Chapter 3 presents the concept of haptic communication and examine the requirements for Tactile Internet to support haptic communication in terms of QoS, QoE and KPIs. Given that haptic teleoperation deals with a human operator controlling an actuator via a communication channel, the authors provide a classification of teleoperation systems based on the type of control system employed. Furthermore, they review the methodologies for teleoperation stability control and haptic data reduction, and then examine the necessary components of the haptic communication infrastructure by taking into account the 5G architecture, 3GPP components as well as the European Telecommunications Standards Institute (ETSI) NFV management and orchestration capabilities.

In Chapter 4, the authors examine how novel 5G concepts such as network slicing, cloud computing, SDN and NFV provide revolutionary features that perfectly fit the communication requirements of smart grid services. The authors provide an overview of future smart grid architecture and the services it is expected to support by categorizing them into data collection and management services. These services support enhanced grid monitoring, as well as control and operation services.

Chapter 5 provides an overview of the evolution of vehicular communication systems towards 5G and show how vehicular applications and services have changed over the years. The authors present a detailed description of the 3GPP specifications for cellular-based vehicle-to-everything (C-V2X) systems from Release 14 through the full standalone 5G system specifications in Release 16 on one hand, and the Dedicated Short-Range Communication (DSRC) on the other hand. Furthermore, a comparison is made between C-V2X and DSRC in terms of channel access, coverage, deployment cost and security among others. The authors also dwell on key 5G technologies that are enablers for V2X services and data dissemination for vehicular communication platforms that could support efficient cloud-based Intelligent Transportation Services (ITS) services through the use of middlewares. Additionally, the chapter analyses the evolution of V2X communication technologies alongside the services they support. The authors assert that there is still lack of insight on the required rollout investments, business models and expected profit for future ITS.

Chapter 6 assesses how the Sparse Code Multiple Access (SCMA) can be used to support high Quality of Experience (QoE) for drivers and passengers of Connected Autonomous Vehicles (CAVs). The authors describe the fundamental principle of SCMA and provide a comprehensive literature review of the advances made in the direction of SCMA as a candidate for 5G air interface, specifically in the case of CAVs. Moreover, the chapter identifies the gaps in the adoption of SCMA in vehicular communication systems.

In Chapter 7, the authors evaluate how URLLC would enhance healthcare delivery and examine two notable use cases – Wireless Tele Surgery (WTS) and Wireless Service Robot (WSR). WTS involves using robotic platforms with audio, video and haptic feedback to perform surgery in remote locations, while WSR deals with robots taking on the role of social caregivers for the sick and elderly in care homes. The authors describe the technical requirements for the implementation of these use cases and 5G key enabling technologies to meet the performance, dependability and security targets for Tele-healthcare. Included in the chapter is a detailed business model and cost analysis that a Tele-healthcare provider would encounter when delivering the corresponding services on a 5G system.

In Chapter 8, the authors elucidate how 5G would disrupt the media and entertainment industry by enabling trends in audio-visual and immersive media. The authors identify the key challenges of this industry, which are capacity, latency and traffic prioritization, and show how 5G would support the services envisioned for the media industry. They start by providing a comprehensive overview of audio-video systems covering wireless audio use case in live production, video compression algorithms, streaming protocols and the requirements for video streaming over mobile networks by considering both practical and theoretical speeds of current mobile networks as well as 5G. With immersive media – augmented reality, virtual reality and 360-degree videos – being considered as the killer applications for 5G, the authors identify numerous use cases as well as the specific QoS requirements for each application and then show how these requirements are within the purview of 5G. In concluding the chapter, the authors posit, "the proliferation of 5G networks will spearhead the adoption of 360°/VR technology in various sectors of the (media) industry and everyday life, and become the catalyst for the expansion of the relevant market".

Chapter 9 presents a realistic mathematical model for Unmanned Aerial Vehicle (UAV)-based cellular coverage that considers a practical directional 3D antenna. The chapter derives analytical expressions for coverage as function of UAV height, beamwidth and coverage radius, and analyses the trade-offs among these factors. Based on in-depth analysis on the effect of altitude and beamwidth on UAV coverage, the authors assert that antenna beamwidth is a more practical design parameter to control coverage, contrary to UAV altitude that is commonly used by researchers. Finally, the authors propose a novel hexagonal packing theory to determine the number of UAVs required to cover a given area.

In Chapter 10, the authors make a case for the use of UAVs to complement cellular systems in order to increase network capacity at certain locations or provide coverage in areas where the existing network cannot. They further present applications of UAVs in cellular networks such as communication in rural areas, data gathering from large-scale wireless sensor deployments, pop-up networks and emergency communications, among others. They then propose an intelligent UAV positioning framework for 5G networks, which leverages on reinforcement learning to determine the best location to deploy the swarm of UAV base stations in an emergency communication scenario.

Chapter 11 provides a comprehensive survey of public safety networks and their historical transition from analogue systems towards 5G. The authors describe the standardization efforts made to support the inevitable convergence of different public safety communication networks to incorporate narrowband and wideband systems as well as the recent trends in communication technologies by pointing out the activities of various standardization bodies such as 3GPP, Open Mobile Alliance, Alliance for Telecommunication Industry Solutions, APCO Global Alliance and Groupe Speciale Mobile Association (GSMA). They further identify connectivity, interoperability and security, among others, as the challenges faced by public safety networks and highlight key 5G technologies to mitigate these challenges.

Finally, in Chapter 12, we make a case for rural mobile data connectivity and the role of government in fast-tracking rural network deployment. Furthermore, we identify and provide insights into some key technologies that would further the gains of 5G in terms of expanding the scope of 5G use cases and applications to new productivity spheres. These technologies include blockchain, terahertz communication, light fidelity (LiFi) and wireless power transfer and energy harvesting and are worth considering in the design of beyond 5G mobile communication systems.

1

Enabling the Verticals of 5G: Network Architecture, Design and Service Optimization

Andy Sutton

1.1 Introduction

5G is often referred to as the next generation of mobile communications technology. However, the potential is much more significant than this; 5G will likely become the future of communications, supporting fixed and mobile access. In addition to enhanced Mobile Broadband (eMBB), 5G will support Ultra-Reliable and Low-Latency Communications (URLLC), also referred to as Mission-Critical Communications, and massive Machine-Type Communications (mMTC) – an evolution of IoT, along with enabling Fixed and Mobile networks Convergence (FMC).

The ITU-R introduces the diverse requirements which 5G will address in recommendation ITU-R M.2083 (09/2015) [1]. These requirements are illustrated in Figure 1.1.

The December 2017 3GPP release of 5G, known as Release 15 phase 1, focused on eMBB use cases through the specification of a new radio interface which connects to an enhanced 4G LTE Evolved Packet Core (EPC) network. The enhanced EPC is referred to as an EPC+ and will likely be implemented using Network Functions Virtualisation infrastructure (NFVi) and therefore be a vEPC+. Because of this a 5G-capable device will connect to an enhanced Release 15 4G LTE radio base station (eNB) for control plane and 4G LTE and/or 5G radio for user plane traffic flows in what is known as Non-Stand-Alone (NSA) mode of operation. This combination of 4G LTE and 5G radio with a 4G EPC+ is what was known originally as an Option 3 architecture, of which Option 3x is the most popular variant. This architecture has been formalized within 3GPP as eUTRA and New Radio Dual Connectivity (EN-DC). While Release 15 doesn't deliver URLLC or mMTC, it doesn't mean there is no support for verticals; eMBB can support a range of use cases and offers a lower level of latency than existing 4G LTE networks. The additional network capacity that will be delivered through an EN-DC architecture will ensure network performance is enhanced while new use cases such as Fixed Wireless Access (FWA) and Hybrid access in which fixed broadband is combined with 5G radio access will enable new business and service opportunities. The 3GPP Release 15 EN-DC architecture is illustrated in Figure 1.2.

To fully exploit 5G for vertical industries it will be necessary to implement the 5G Next Generation Core (NGC) network along with NR. This delivers a full 5G end-to-end

Enabling 5G Communication Systems to Support Vertical Industries, First Edition.
Edited by Muhammad Ali Imran, Yusuf Abdulrahman Sambo and Qammer H. Abbasi.
© 2019 John Wiley & Sons Ltd. Published 2019 by John Wiley & Sons Ltd.

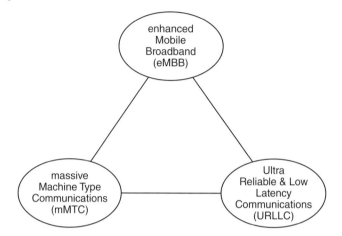

Figure 1.1 Diverse network requirements of 5G.

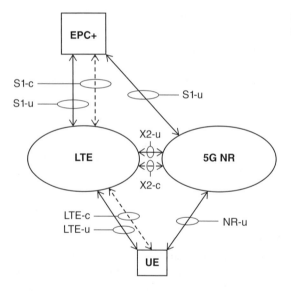

Figure 1.2 3GPP EN-DC network architecture.

network as specified by 3GPP. There is a Stand-Alone (SA) version of 5G once the NGC becomes available. However, many networks will implement NGC is what was origi-nally known as an Option 7 network architecture (Option 7x being the most common variant) in which the radio control plane will remain on 4G LTE due to the probability of superior 4G LTE radio coverage in the early days of 5G rollout while the user plane will be on 5G radio and/or 4G LTE. This dual connectivity solution with a 5G NGC has been specified by 3GPP as Next Generation eUTRA and New Radio Dual Connectivity (NGEN-DC). Once 5G coverage is equivalent to or better than 4G LTE the control plane can be switched to the 5G radio base station, known as a gNB. It is anticipated that in the fullness of time mobile operators will re-farm their 4G LTE radio spectrum to 5G NR; however, this is expected to take many years, considering how much spectrum is still being utilized for older 2G GSM/GPRS and 3G UMTS/HSPA networks.

1.2 Use Cases

A 5G end-to-end network, with or without dual connectivity with 4G LTE, will support a wide range of use cases. The Next Generation Mobile Networks (NGMN) Alliance's 5G white paper [2] presents the 5G services as eight use case families and highlights a single example of each use case family, noting that there are many other use cases within each family which will address a wide range of connectivity requirements. Example use cases are illustrated in Figure 1.3.

The true diversity of capability requirements which 5G must address are illustrated in Figure 1.3, from eMBB use cases to those requiring low-latency and/or ultra-reliable communications along with use cases which will require mMTC. The forecast for growth of the Internet of Things has led to the ITU-R specifying the need for mMTC to support up to 1,000,000 devices per km^2.

eMBB will address broadband data access in dense urban areas through the provision of very high area capacity density while also addressing the opposite extreme of broadband access everywhere, which targets 100% geographical coverage. Broadband on trains is a key use case for higher user mobility, which will require a reliable eMBB connection albeit reliability at such high speeds will bring unique challenges. Other use cases for higher user mobility include mobile broadband connectivity to aircraft, as demonstrated with the implementation of the European Aviation Network (EAN) [3] and future use cases involving drones. The massive Internet of Things will make use of mMTC, likely an evolution of 4G LTE Narrow-band – IoT (NB-IoT) technology to provide the capability to support up to 1 million devices per square kilometre. Use case families, extreme real-time communications, lifeline communications and ultra-reliable communications will really stretch the capabilities of 5G. Capabilities of eMBB and mMTC will be used for certain use cases within these families. However, the use of URLLC, be this ultra-reliable and/or low-latency communications, will consume significant resources from the network, effectively shrinking radio coverage and reducing capacity as reliability and/or low-latency targets become more extreme [4]. The last use

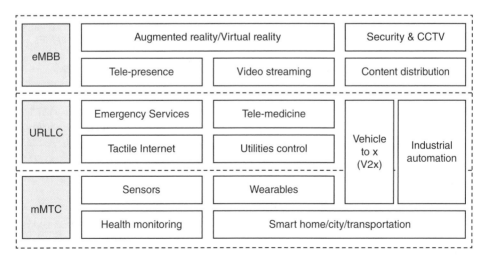

Figure 1.3 Example use cases for 5G network capabilities.

case family is broadcast-like services; an evolution of 4G LTE enhanced Multimedia Broadcast Multicast Service (eMBMS) coupled with IP multicast within the transmission network, this will enable efficient distribution of mass market real-time content such as live sports or news and support multicast use cases from retailers and other verticals.

1.3   5G Network Architecture

A 5G end-to-end network supports subscriber data management, control plane functions and user plane functions. Since the early days of Global System for Mobile Communications (GSM) and then General Packet Radio Service (GPRS), the mobile eco-system has been familiar with logical representations of mobile network architectures. These diagrams take the form of functional blocks and the interfaces between them, officially known as reference points. Figure 1.4 presents this view of the 3GPP 5G network, referred to as "reference point representation".

The reference points or interfaces, which will be known as interfaces for the remainder of this chapter, start with the letter "N". Originally these were designated "NG" for next generation. However, recently the term has been shortened to simply read "N". The functional blocks are split between control plane and user plane functions, with the control plane further split between subscriber management functions and control plane functions. The subscriber management functions consist of the Authentication Server Function (AUSF) and Unified Data Management (UDM), an evolution of the Home Subscriber Service (HSS). The control plane function consists of a core Access and Mobility management Function (AMF), a Session Management Function (SMF), Policy Control Function (PCF), Application Function (AF) and Network Slice Selection Function (NSSF). The NSSF is responsible for selecting which core network instance is to accommodate the service request from a User Equipment (UE) by considering the UE's subscription and any specific parameters. The user plane path starts with the UE, which

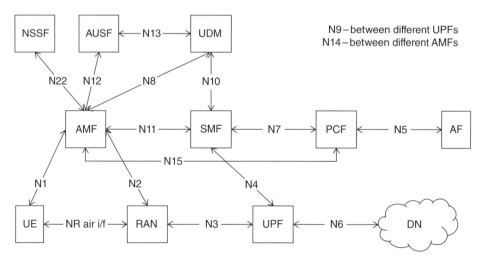

Figure 1.4 5G network architecture – reference point representation.

may be a smartphone or a new form factor terminal, possibly fixed rather than mobile. The UE connects via the Radio Access Network (RAN) to the User Plane Function (UPF) and on to a Data Network (DN). The DN may be the public Internet, a corporate intranet or an internal services function within the mobile network operator's core (including content distribution networks).

The NR air interface downlink waveform is Cyclic Prefix-Orthogonal Frequency Division Multiplex (CP-OFDM) access while the uplink can be either CP-OFDM or Discrete Fourier Transform-spread-Orthogonal Frequency Division Multiple (DFT-OFDM) access, the uplink mechanism being selected by the network based on use case. The UE connects to the RAN via the air interface, which also carries the N1 interface, which in previous iterations of 3GPP technologies has been known as the non-access stratum. The N1 interface is a secure peer to peer control plane communication between the UE and core AMF. The N2 and N3 interfaces comprise what is commonly known as mobile backhaul between the RAN and the core network although, as we'll discuss shortly, this isn't as simple in reality as the illustration in Figures 1.4 and 1.5 suggest. The N6 interface provides connectivity between the UPF and any internal or external networks or service platforms. The N6 interface will include connectivity to the public Internet and will therefore contain the necessary Internet-facing firewalls and other smarts associated with the evolution of the Gi/SGi LAN environment. The Gi/SGi LAN environment has evolved from GPRS through UMTS and LTE to provide a range of capabilities in support of mobile data network operation, including features such as Transmission Control Protocol (TCP) optimization, deep packet inspection and network address translation.

In addition to the familiar logical network diagram with defined interfaces, 3GPP has introduced an alternative view of the 5G network architecture which is known as Service-Based Architecture (SBA). SBA takes advantage of recent developments in Network Functions Virtualisation and Software Defined Networking to propose a network based on virtualized infrastructure. This architecture will leverage service-based interactions between control plane functions as necessary. The solution will sit on

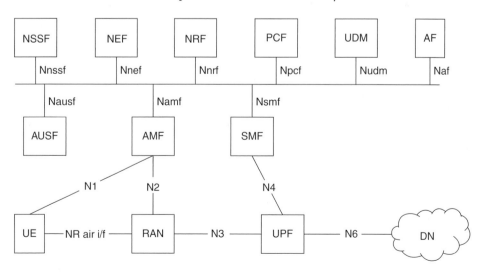

Figure 1.5 5G service-based architecture.

common computer hardware and call upon resources as required to manage demand at any instance. The use of SBA does not mandate a centralized solution; distributed computing could be implemented if appropriate. The SBA is illustrated in Figure 1.5.

3GPP states a number of principles and concepts for SBA [5] (not all are exclusive to SBA), including:

- separate control plane functions from user plane functions, allowing independent scalability, evolution and flexible deployment;
- modularize the functions design to enable flexible and efficient network slicing;
- wherever possible, define procedures (the interactions between network functions) as services so that their re-use is possible;
- enable each network function to interact with other network functions directly, if required;
- minimize the dependencies between the access network and core network; this will enable different access types such as fixed broadband and Wi-Fi;
- support a unified authentication function;
- support stateless network functions such that the compute resource is decoupled from the storage resource;
- support concurrent access to local and centralized services; this will enable support for low-latency services along with access to local data networks. To facilitate this, user plane functions can be deployed much closer to the access network;
- support roaming with both home network routed traffic and local break-out traffic in the visited network.

The SBA introduces a couple of functions that didn't exist in the traditional logical interface-based architecture representation; these are the Network Repository Function (NRF) and Network Exposure Function (NEF). The NRF provides control plane network functions with a mechanism to register and discover functionality so that next generation control plane network functions can discover each other and communicate directly without making messages pass through a message interconnect function. The NEF receives information from other network functions (based on exposed capabilities of other network functions). It may store the received information as structured data using a standardized interface to a data storage network function. The stored information can be re-exposed by the NEF to other network functions and used for other purposes such as analytics. A practical example of use of the NEF is to aid the establishment of an application server-initiated communication with a UE where no existing data connection exists.

In the high-level network architecture illustrated in Figures 1.4 and 1.5, the RAN is represented as a single functional entity whereas in reality the realization of a 5G RAN is not so straightforward. In GSM/GPRS and UMTS there is a network controller which provided an interface between the radio access network and the core network. This network controller hid a lot of signalling from the core, particularly in UMTS, and managed a range of complex RAN functions. In LTE there is no network controller, the RAN manages a range of mobility management and radio optimization activities between evolved Node Bs via the X2 interface. 5G effectively introduces a centralized RAN node albeit not a network controller as such. The 5G radio base station, known as a next Generation Node B (gNB) is split into three entities: a radio unit (RU), a gNB-Distributed Unit (gNB-DU, often shortened to DU) and a gNB-Centralized Unit (gNB-CU, often shortened to CU). The protocol layer interface at which this split will occur has been the topic of much debate in 3GPP and throughout the wider industry.

1.4 RAN Functional Decomposition

3GPP used the RAN protocol model illustrated in Figure 1.6 [6] to discuss the functional split which should be implemented in 5G. Note that this protocol model is based on LTE as this was all that was known at the time although this doesn't differ significantly from 5G NR. The same terms are used to describe the radio interface protocol layers although there have been some minor movements of functional sub-entities. Additionally, a new protocol known as Service Data Adaptation Protocol (SDAP) has been introduced to the NR user plane to handle a flow-based Quality of Service (QoS) framework in the RAN, such as mapping between QoS flow and a data radio bearer, and QoS flow ID marking.

Reading Figure 1.6 from left to right, Radio Resource Control (RRC) resides in the control plane while the data is in the user plane. As discussed above, SDAP will be inserted between data and the Packet Data Convergence Protocol (PDCP) for a 3GPP standards-compliant 5G NR view of the protocol stack. Functions of PDCP include; IP header compression and decompression along with ciphering and deciphering (encryption of the data over the radio interface). PDCP feeds down the stack to the Radio Link Control (RLC) layer. RLC functions include: error correction with Automatic Repeat request (ARQ), concatenation and segmentation, in sequence delivery and protocol error handing. Moving down the stack from RLC to the Medium Access Control (MAC) layer we find the following functions; multiplexing and de-multiplexing, measurement reports to RRC layer, hybrid ARQ error correction, scheduling and transport format selection. The physical layer is responsible for a range of layer 1 functions, including: modulation and demodulation of physical channels, frequency and time synchronization, power weighting of physical channels, MIMO antenna processing and RF processing.

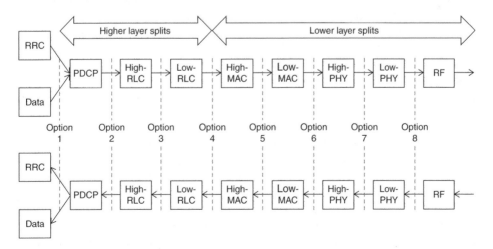

Figure 1.6 RAN protocol architecture as discussed in 3GPP TR 38.801.

Cell site–DU co-located

Centralised aggregation
site

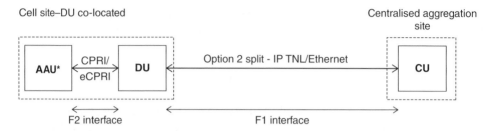

*AAU illustrated, actual implementation could be AAU or passive antenna with RRU

Figure 1.7 RAN functional split Option 2 DU – CU split architecture.

After much debate 3GPP agreed on an Option 2 functional split, meaning that PDCP and therefore everything above this layer will reside in the CU while RLC and everything below it will reside in the DU. This is known as a higher layer split given its location within the protocol stack. The interface between the CU and DU has been designated "F1" [7] as shown in Figure 1.7; this will be supported by an IP transport network layer which will be carried over an underlying Carrier Ethernet network as specified by the Metro Ethernet Forum [8]. Given the chosen location of the functional split there are no exacting latency requirements on the F1 interface; in fact, it's likely that the latency constraints applied to the F1 interface will be derived from the target service-based latency. The data rate required for the Option 2 F1 interface is very similar to that of traditional backhaul (LTE S1 interface as a reference) [9] for a given amount of spectrum multiplied by the improved spectral efficiency of NR. A higher layer split doesn't enable the real-time coordinated scheduling benefits of a lower layer split. However, benefits of the Option 2 split include load-balancing between adjacent cells and optimization of the X2 interface which results in much lower latency for X2 control and user plane transmissions.

5G will build on the trend towards separate BaseBand Unit (BBU) and Remote Radio Unit (RRU) as increasingly deployed in today's mobile networks. The common definition of Cloud or Centralized RAN (C-RAN) refers to a common centralized BBU supporting multiple cell across a geographical area, the 4G LTE implementation is based on an Option 8 Common Public Radio Interface (CPRI) split which is the extended over a wide area via a low-latency point to point transmission network, typically over fibre optic cables although wireless solutions can be used in certain scenarios. C-RAN is not common in European networks. However, the same split architecture is implemented but the CPRI interface is constrained to connectivity within a cabinet or between a cabinet mounted BBU and local tower mounted RRU.

5G introduces massive MIMO antenna systems which integrate the RRU functionality within the antenna unit; these are known as Active Antennas Units (AAU). While the interface between the current 4G LTE BBU and RRU spectral efficiency is set to 2.25 bits per is based on the CPRI protocol in most implementations, recent developments by the CPRI group, which consists of most of the major mobile RAN manufacturers, has resulted in an alternative lower layer split known as evolved CPRI (eCPRI) [10]. The initial implementation of eCPRI maps to an Option 7 split. It is likely that both CPRI and eCPRI interfaces will be supported between DU and AAU as there are pros and cons to

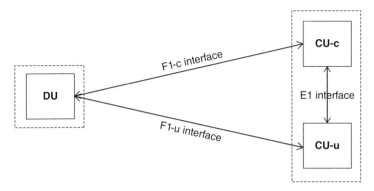

Figure 1.8 F1 interface with split CU control and user plane entities.

both approaches, strategies will differ between equipment vendors and between mobile network operators.

The interface between the DU and AAU is often referred to as the F2 interface (although this isn't a formal 3GPP term, it may be adopted in the future). This interface may be local to the cell site or could be extended to form a more coordinated RAN. The challenge with extending this interface across a wide area is the exacting performance requirements in terms of ultra-low latency and extremely high data transmission rates, particularly in the case of CPRI; eCPRI does benefit from some compression to reduce the data rate. This is the factor which has constrained the use of C-RAN in some markets, particularly Europe, while the abundance of dark fibre in markets in Asia has led to the mass adoption of C-RAN in urban areas. It should be noted that it is possible and quite likely that some operators will co-locate the DU and CU at the cell sites and therefore effectively deploy a 5G distributed RAN (D-RAN), as used extensively today with 2G, 3G and 4G networks, as an alternative to C-RAN.

The theme of decoupling control and user plane is central to 5G network architecture development and therefore it is natural that this should be considered for the F1 interface. The F1 interface is split into control and user plane interfaces which are known respectively as F1-c and F1-u as shown in Figure 1.8. The CU itself is also split into two functional entities; these could exist on the same hardware, on separate hardware on the same site or on separate hardware across different physical site locations. To connect the decomposed CU a new interface, designed E1, has been defined within 3GPP. The E1 interface (not to be confused with the legacy E1 (2.048Mbit/s) transmission interface) connects the CU-c and CU-u functional entities. The CU in its entirety can be built using virtualized infrastructure, as can the majority of the 5G network, the noticeable exception being certain radio frequency functions such as power amplifiers and antennas.

1.5 Designing a 5G Network

As already discussed, 5G will be deployed in parallel with 4G LTE in most practical deployments, some Fixed Wireless Access (FWA) deployment are likely in certain markets in isolation of 4G integration. New use cases for 5G for vertical industries will rely on mainstream mobile network operator integrated solutions and bespoke deployment scenarios to address specific needs, these could be deployed by a mobile operator in licensed and/or unlicensed spectrum or could be deployed by a specialist provider using unlicensed or CBRS spectrum. A 5G version of MulteFire [11] is most likely, which will enable 5G NR in unlicensed spectrum, including the 60 GHz unlicensed band.

With an understanding of the 3GPP 5G logical network architecture, the products and roadmaps of network equipment vendors and an appreciation of the services to be offered it is possible to start designing a 5G network architecture. The following example is based on the BT network in the UK. However, the principles are generic and can be applied to other networks. The example can be applied to the deployment of a 5G Next Generation Core (NGC) network in support of a NGEN-DC architecture (formally known as Option 7x). However, the principles are the same for any 5G network architecture, including standalone 5G deployment.

To start a network architecture design, it is essential to understand the existing network on which the new technology will be overlaid. Many mobile networks will have legacy Radio Access Technologies (RATs) deployed and will therefore have an established base of cellular radio sites (macro and micro-cells). Mobile backhaul [9] will be in place to support the existing technologies and this will terminate at the edge of the mobile core network, possibly via intermediate transmission sites. The core network will likely consist of a hierarchy of sites which connect to service platforms, Internet peering and transit along with voice telephony interconnects for circuits switched and Voice over IP (VoIP) services.

The example network used in the following explanation consists of cell sites, access nodes (could be telephone exchange sites in the case of an incumbent operator), aggregation (or metro) nodes and core network sites along with connectivity to external network via Internet Peering Points (IPP). The sites typically serve as lower hierarchical functions along with their senior hierarchy, therefore a core site will also act as an aggregation node and access node while an aggregation node will also act as an access node. Cell sites are connected either by optical fibre-based connectivity or point to point wireless backhaul, this could be microwave or millimetre-wave radio systems. Other networks may use point to multi-point or multi-point to multi-point (mesh)-based wireless systems; these systems will become a greater consideration as network densification is implemented with small cells. Naming conventions for cell site types vary considerably within the industry. Macro-cells and micro-cells may well use the same equipment; however, they will likely vary in their geographical coverage and configurations, including transmit output power. As 4G LTE evolves and 5G is introduced, we will witness an increasing deployment of external and internal pico-cells, often referred to as small cells. These will be smaller still from a coverage perspective. However, they may have significant capacity in the future as ever-higher frequency spectrum will be deployed and eventually an area of contiguous small-cell coverage will exist to significantly increase area capacity density and enhance user experience, resulting in ever lower latency and higher peak and average data rates. A typical network structure is illustrated in Figure 1.9.

To understand how best to place 5G network components across the end-to-end network requires an analysis of the products and service requirements. Initial 5G networks will address the eMBB use case and therefore most subscribers will either be accessing the Internet or will be corporate users accessing their private intranets via dedicated Access Point Names (APN). In these scenarios the content (service) a subscriber wishes to consume is located on an external network or potentially at a centralized on-net

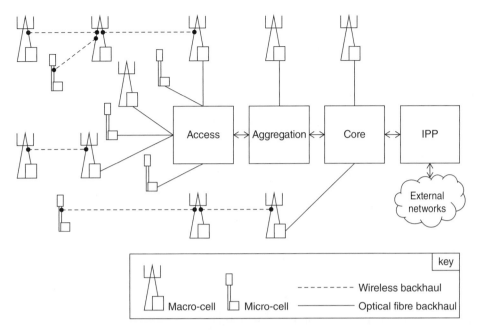

Figure 1.9 Mobile network structure and site types.

Content Distribution Network (CDN). Either way, the communications will traverse the end-to-end network to enable the service. The eMBB use case will benefit from the improved peak and average data rates that 5G will offer, possibly working in NGEN-DC mode with 4G LTE providing the wide-area coverage and 5G enabling a speed boost, and as more spectrum is deployed/re-farmed for 5G NR then a move to a 5G controlled dual connectivity mode or standalone architecture will be possible. The allocation of user plane capacity in dual connectivity mode between 5G NR and 4G LTE will be determined by the vendor-specific scheduling algorithms on the base stations. However, it is possible for a user to consume resources from the 5G NR without needing additional resources from 4G LTE; this would likely offer a lower latency than a dual connected user plane.

1.6 Network Latency

Network latency in 5G is receiving much attention, with figures of 1 ms latency grabbing many headlines. The reality and, indeed, the network latency requirements are quite different to this. The ITU-R within their IMT-2020 requirements document and 3GPP within their 5G target performance specification have specified the latency targets shown in Table 1.1 for one-way transmission over the radio network.

The NR TDD slot format will ultimately constrain the minimum latency which can be achieved over the radio interface. Regulatory considerations with regard to slot format alignment and synchronization between competing mobile network operators (sharing the same NR TDD band) may constrain the choices of slot format to a subset of those specified within 3GPP standards.

Table 1.1 ITU-R and 3GPP one-way radio interface latency targets [12].

Organization	URLLC	eMBB
ITU-R (IMT-2020)	1 ms	4 ms
3GPP (5G)	0.5 ms	4 ms

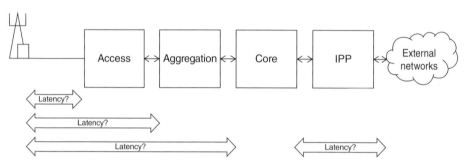

Figure 1.10 High-level architecture and latency considerations.

Understanding end-to-end network latency is an essential input to any node placement within a network. Taking the example network as illustrated in Figure 1.10, we can identify many components within the network latency budget. There will be a processing delay within the UE and base station which is often included within the radio interface latency budget. After this the transmission circuit (mobile backhaul) must be considered; a typical Internet access data service will transit several sites within the operators IP transport network and likely pass through various network components, including: routers, switches, IP security gateways, mobile core network and service LAN (Gi/SGi/N6 interface), along with any Internet peering or transit.

To reduce latency for eMBB services that wish to access content on the public Internet or a private intranet means minimizing the end-to-end latency within the network. This can be achieved by minimizing the number of transmission hops and network components within the path. In addition, all transmission paths should point towards the Internet peering points to minimize unnecessary physical transmission distance. For eMBB access to distributed content, the question is: what level of distribution is appropriate? This question is important when considering the design of services for vertical industries; quite often these will have specific Service Level Agreements (SLA) against performance parameters, such as availability, latency and throughput. Existing mobile networks are increasing making use of on-net CDN to enhance user experience; often these are provided by a third-party content provider or aggregator and sit on core network sites. These sites are quite deep in the network. However, they avoid the need for connectivity via peering points and access to external servers, thus reducing latency to enhance the user experience while reducing costs for the network operator.

An analysis of network latency across the end-to-end network is possible through network modelling. However, it is always best to validate any modelling simulations with a range of practical field measurements which will inform the development of a latency

Table 1.2 Network latency figures for cell site to network node (source: BT [13]).

Node	One-way latency
Access (1000 sites)	0.6 ms
Aggregation (100 sites)	1.2 ms
Core (10 sites)	4.2 ms

model. Based on the network structure in Figure 1.10, results for the actual latency from a cell site to the access node, aggregation node and core node were produced. This data from BT [13] is based on an analysis of 8000 cell sites deployed within the UK with fibre backhaul connectivity. The model utilizes 1000 access nodes, 100 aggregation nodes and 10 core sites. This analysis is vital because deploying capability in 10 sites is cheaper than 100 sites while 100 sites is significantly cheaper than 1000 sites, so an appropriate network architecture must be developed to support the business opportunities.

The figures in Table 1.2 represent the 95th percentile and as such many sites will have lower transmission latency. However, this does provide a useful starting point for building out a 5G network architecture. With these figures and those from 3GPP (which appear in Table 1.1) it is possible to produce a high-level approximation of network latency to inform the network architecture development and any phasing of subsequent network evolution, including core user plane distribution.

5G network elements or components are specified by 3GPP, though they may be implemented using NFVi and as such should be considered as logical entities which may or may not be linked to specific hardware. This is an evolving concept in telecommunications, though it has become quite common in IT data centre environments. The optimization of control and user plane may be necessary depending on use case. A UE may be a smartphone, tablet or mobile Wi-Fi hotspot. However, equally it may be a broadband residential gateway with a mains power supply; this will likely impact the uplink performance of the device and if connected to an external antenna, the overall radio link budget will be improved. Alternatively, the UE may be an Internet of Things (IoT) device which again may be battery-powered, as in the case of a remote sensor or wearable, or could be mains-powered in the case of a piece of industrial machinery. Considering use cases within the automotive industry and the role 5G could play in Connected and Autonomous Vehicles (CAV) is interesting, as vehicles have a high-capacity power system on board which could support not only the functions of a UE but also network function such as relays and computing/storage.

1.7 5G Network Architecture Design

Initial 5G-based eMBB services will likely access content from the public Internet, a private intranet or service platforms sitting within the mobile operator's core network. As the content is sitting deep in the network or on an external network, there are practical limitations on how low the latency can be. Due to the location of the content to be served, there is little benefit to be realized from a significant amount of core network user

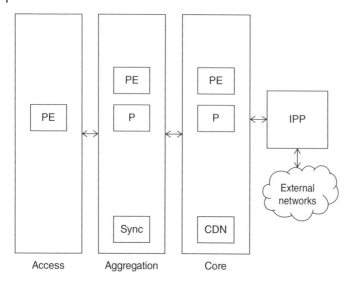

Figure 1.11 Base network architecture ready for 5G overlay.

plane distribution. Deploying the 5G UPF to more centralized locations, likely the same locations as the control plane functions will minimize costs and operational overheads necessary to serve the eMBB use case.

The network architecture diagram of Figure 1.10 can be expanded to review key network nodes as illustrated in Figure 1.11. Network access is facilitated via a PE router, this node is often known by other functional names within network operators; for example, BT refers to the PE router as a Multi-Service Edge (MSE) router which highlights the capability of the platform to support multiple traffic types and services. This node also provides fixed network Broadband Network Gateway (BNG) services on a subset of PE/MSE nodes, as a result of this the fixed broadband network facing interface is the same as found on a mobile Gi/SGi/N6. This interface exposes the user IP stack and can therefore be used for service injection.

Alongside the IP/MPLS (Internet Protocol/Multi-Protocol Label Switching) network components of PE and P routers there's a Sync platform and CDN illustrated in Figure 1.11. The Sync platform can provide frequency and phase synchronization to meet the requirements of the 5G TDD radio interface while the CDN provides access to content from on-net (within the same IP autonomous system as the mobile core network) servers, saving the need to transit or directly peer via the Internet. CDN can support fixed and mobile customers. The example model used to explore the development of a 5G network architecture consist of 10 core sites to 100 aggregation sites and 1000 access sites, each of which can support directly connected cell sites supporting 4G LTE and 5G radio access nodes (legacy 2G and 3G services may also be supported).

As 5G radio coverage increases to match or exceed that of 4G LTE, the control plane will be switched to the 5G gNB, at which point the network architecture becomes an Option 4 network architecture, and this continues to support dual connectivity for the user plane. The alternative to a dual connectivity model is a fully standalone 5G network, known as Option 2 network architecture, which will be achieved once all 4G LTE

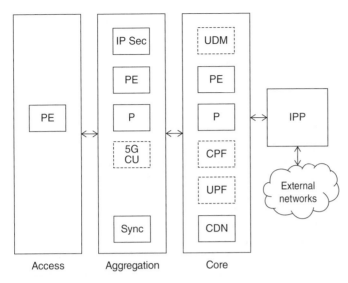

Figure 1.12 Initial 5G network overlay for eMBB use case.

spectrum is re-farmed to 5G or in the case of FWA, when 5G is used without dual connectivity as a standalone solution.

Figure 1.12 introduces 5G NGC components to the core network sites, likely just a subset of the 10 sites initially. The UDM will probably remain quite centralized, being backed up across several sites while the control and user plane functions will start on a few sites and scale as necessary to the maximum of 10 sites (in this model). The implementation of the CPF and UPF components of the NGC will utilize NFVi and therefore consist of a generic hardware platform to provide compute capability with specialist software providing the actual 5G functions. It is highly likely that early implementations will run in Virtual Machines (VMs) while in time this may evolve to a container-based approach. The centralized UPF provides access to on-net CDN and external networks as required, the use of dedicated APNs will enable verticals to build solutions within their secure intranets while allowing access from mobile devices on the mobile operator's network. The list of devices will include conventional smartphones and tablets. However, increasingly we'll see devices ranging from broadband residential gateways to telemetry and sensor networks, some will utilize radio services based on NB-IoT, strictly a 4G LTE technology, in parallel with 5G services.

The CU shown in the aggregation site is a likely starting point for a cloudified radio access network, whether an operator implements a fully distributed RAN or one of several cloud/centralized RAN options from day one will be determined by several factors. Different RAN vendors have taken different approaches to their 5G RAN product line. Some start with combined DU and CU, therefore supporting a single-site solution or traditional CRAN with a CPRI or eCPRI interface between RU and BBU (DU/CU). Other vendors have implemented a split DU/CU in accordance with the 3GPP F1 interface specifications. This enables an Option 2 split as previously illustrated in Figure 1.7. The F1 interface is implemented as a 3GPP IP Transport Network Layer (IP TNL) and as such will need to be secured to ensure the integrity of the interface. Integrity protection can be provided using IP Security Authentication Header (AH) via a security gateway, identified as a non-5G specific component in Figure 1.12 and labelled as "IP Sec". As the CU contains the PDCP layer, the connection will be encrypted. However, additional encryption could be applied using IP Security Encapsulating Security Payload (ESP) feature. If there is no CU, therefore implying a DRAN, ESP will be essential alongside AH as PDCP will reside on the cell site.

Figure 1.12 proposes a CU in up to 100 aggregation sites, this would be the start of a cloudified RAN and serve to facilitate a level of coordination between adjacent gNB sites. The F1 interface-based split is non-real-time and therefore won't facilitate the tight coordination of a true-CRAN deployment. Advantages of an F1 interface split include load-balancing between adjacent cell sites and X2 interface optimization, particularly useful once spectrum is re-farmed from 4G LTE to 5G or in a truly standalone 5G deployment. Prior to this the dual connectivity solution will require X2 between eNB and gNB. If a F1 interface split is selected, a techno-economic analysis is required to determine how many network sites will host CU capability. Increasing the number of sites will likely reduce the latency on the F1 interface whereas reducing the number of sites will reduce the overall cost of deploying the solution. Reducing F1 interface latency may initially seem like a good idea. However, it isn't necessarily a critical factor for non-real-time RAN protocol layers while the end user latency will be determined from where the content is served, in this case from the core sites or external networks. Placing the CU in the aggregation layer does offer economic benefits over the access sites, and based on the latency analysis in Table 1.3 seems an optimal location to support low-latency services in the future.

As 5G services evolve with the introduction of URLLC and mMTC, there will be significant growth in the number of vertical use cases which can be addressed. URLLC brings a mini-slot structure to the 5G radio interface which allows for ultra-low latency transmission when implemented with higher numerologies and a suitable uplink access scheme. Supporting lower latency and ultra-reliable communications will have an impact on overall network design; from a reduced link budget resulting in effectively less coverage in a given radio band/configuration and reduced capacity due to the need to pre-schedule resources, through to the need for diverse transmission paths and battery back-up on cell sites. Network architecture requirements for low-latency communications will be influenced by the analysis presented in Tables 1.2 and 1.3. From this analysis it is possible to consider the placement of distributed UPF and associated service platforms; these could be CDN or Multi-access Edge Computing (MEC). In the case of MEC, the platform itself could act as the NFVi on which network components are implemented, along with acting as a service platform for CDN, local net-apps hosting or compute as a service use cases.

Table 1.3 Estimated 5G latency performance (source: BT [13]).

Node =	Access	Aggregation	Core
Number of sites	1000	100	10
One-way transmission latency	0.6 ms	1.2 ms	4.2 ms
Estimated 5G latency (RTT) eMBB	9.2 ms	10.4 ms	16.4 ms
Estimated 5G latency (RTT) URLLC	2.2 ms	3.4 ms	9.4 ms

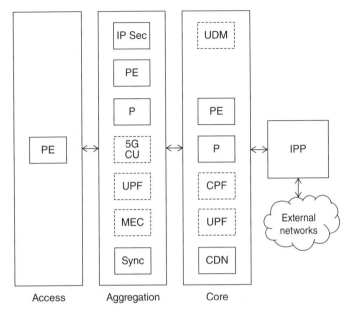

Figure 1.13 User plane and service platform distribution to aggregation sites.

Taking the base model and overlaying an initial phase of user plane distribution would enable lower latency network-based services. This is illustrated in Figure 1.13.

Figure 1.13 illustrates a level of distribution of the core user plane, the UPF has been distributed to the aggregation layer therefore increasing the number of UPF instances from up to 10 to a maximum of 100. Note that not all aggregation sites need to be equipped with UPF and service platforms, a network planning exercise informed by a techno-economic analysis will be required to determine the correct number of sites to be enabled for distributed user plane functions.

The concept of network slicing is enabled through the 5G NGC and as such a user may access different slices which terminate on different UPFs. It is the role of the CPF, in particular the AMF, SMF and NSSF, to determine which UPF a particular session is connected to for a given network slice or service function. Example use cases for URLLC and mMTC are listed in Table 1.4.

Table 1.4 URLLC and mMTC use cases.

URLLC	mMTC
Emergency services critical communications	Smart meters and sensors
Industrial applications and control	Fleet and vehicle management
Remote medical services	Smart agriculture
Factory automation	Logistics
AR/VR for real-time collaboration	Tracking
Support of connected and autonomous vehicles	Smart cities and roads

The architecture illustrated in Figure 1.13 will require a wider distribution of NFVi capability, which for reasons of scale may differ from the approach taken in the core network sites. As equipment is pushed towards the edge of the network, the amount of traffic/services managed at a particular location will be lower than at a more centralized core network site, and as such any overheads to do with running cloud operating systems etc. will form a greater percentage of the overall cost base. Once the distributed user plane has been implemented, it will be possible to support lower latency services which are hosted locally. Any service which requires access to centralized resources or an external network can continue to be served from a centralized UPF. In the example model the aggregation node could serve latency of approx. 10.4 ms RTT for traffic accessing the network via an eMBB radio interface format, while traffic using the URLLC radio interface format could be served with latency of approx. 3.4 ms (noting previous comments about the impact of URLLC radio on coverage and capacity). URLLC could be deployed in a dedicated spectrum band to address specific business opportunities. An example use case which may require national URLLC coverage is vehicle to infrastructure (V2x) connectivity for critical safety systems. These figures are based on the analysis presented in Table 1.3 and are approximations, as in reality there would be other considerations such as processing delay in the UE and network components along with access time to any content/compute resources on the MEC platform. This does, however, explain the principle of user plane distribution and illustrates that low-latency services of approx. 10 ms and <5 ms could be served from network-based resources. This also enables significant financial synergies to be realized when compared with the deployment of MEC to every radio base station, it also makes service continuity much easier as it allows for a level of mobility if required. <10 ms would be available to many users and could therefore support applications such as AR and VR while <5 ms could support lower latency requirements for certain critical use cases.

To reduce network-based latency further requires an additional level of user plane distribution, moving from up to 100 sites to a maximum of 1000 sites. Distributing network functionality to the access sites will increase costs considerably so as with the initial phase of user plane distribution previously discussed, a subset of sites could be selected based on specific business opportunities (if the service is local or regionally based then the access sites enablement is only required to cover this geographical area; other areas may continue to access services via the aggregation site infrastructure). Moving the UPF to the access sites would incur additional costs as IP Security gateways would be required at the edge of the network; these tend to be specialized hardware-based platforms currently. However, it is anticipated that virtualized IP Security gateway functionality will be available at a similar performance specification to the dedicated hardware solutions by 2020. To terminate the user plane at the access layer would require CU functionality moving from the aggregation sites to the access sites too, in addition to the UPF and MEC requirements. Overall this is a significant investment in network infrastructure. However, as can be seen from Table 1.3, this would enable approximately 9.2 ms latency for eMBB use cases and approx. 2.2 ms for URLLC use cases. While the latency performance offered by deploying capability at the access site layer is an improvement on that of the aggregation layer, it is essential to review the service portfolio to understand the services and associated revenue opportunities that can be addresses by the 1.2 ms reduction in latency. Figure 1.14 illustrates the distributed user plane and service platforms architecture deployed to the access site layer.

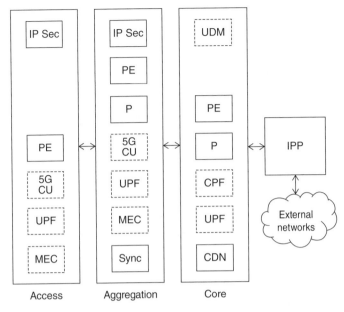

Figure 1.14 User plane and service platform distribution to access sites.

The access nodes could serve an additional function in support of RAN functional decomposition. If suitable transmission was available between the cell site and access site, an eCPRI-based fronthaul solution could be implemented. This enables a full CRAN architecture, which would enhance performance of the radio interface in many respects when compared with a traditional DRAN. Having access to fibre-based transmission is essential, as the eCPRI would consume one or more optical wavelengths on a Wavelength Division Multiplexing (WDM) access system. If this solution was implemented, the DU would move from a number of individual cell sites to an access site where multiple RU from multiple cell sites would terminate, enabling coordination between adjacent cells and removing the need for an external X2 interface. If NGEN-DC operation is in use then having a DRAN for 4G LTE and CRAN for 5G NR will present some challenges. This solution lends itself to having CRAN for both RATs or having a 5G standalone network architecture.

To meet the ultra-low latency targets of 1 to 2 ms will require a local and therefore likely bespoke solution to address a particular use case. It is anticipated that this requirement will not be network-wide and may be restricted to solutions such as industrial automation and local remote surgery (in which the patient and surgeon are in the same building) in the first instance. Other ultra-low latency use cases will be addressed by direct device to device communications, for example use cases in automotive platooning.

Figure 1.15 illustrates the cell site-based ultra-low latency service delivery solution in which the complete chain of network components and service platform is located on a cell site. This may be a standard wide-area cell site or more likely a bespoke deployment to address the specific use case (an end-to-end 5G network solution built within a factory etc.). The RU could be integrated into an AAU or a standalone unit, and the DU, CU, UPF

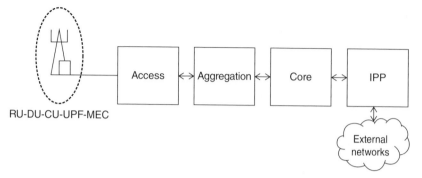

RU-DU-CU-UPF-MEC

Figure 1.15 Service enabled cell site for ultra-low latency use cases.

and MEC could be implemented across multiple virtualized infrastructures or could sit on a common infrastructure and run as separate VM/VNF.

1.8 Summary

5G technology will address a wide range of use cases, far more than any previous generation of mobile communications technology. These use cases include enhanced mobile broadband, ultra-reliable and low-latency communications and massive machine-type communications. A key driver for 5G is to serve new and evolving use cases from vertical industries while continuing to enhance the wide-area mobile broadband experience for all subscribers. The introduction of higher-order and massive MIMO antennas systems with beamforming will deliver ever-higher throughput and increase cell area capacity density through the enhancement of spatial multiplexing. 5G will operate in a much larger range of frequency bands which extend from the sub-6GHz bands which have been common to date with cellular mobile systems, to higher centimetre- and millimetre-wave frequency bands. The higher frequency bands enable much wider radio channels, which will support peak data rates in the tens of gigabits per second, while coupling these wide channels with suitable uplink access mechanisms and mini-slot structures will deliver ultra-low latency. There is not an industry or business sector that will not be impacted by the introduction of 5G in support of an increasingly connected and automated digital workplace. The way we communicate, work, travel and spend leisure time will continue to evolve at a rapid rate, with new and innovative services such as AR and VR, enhanced mobile broadband on train and aircraft and connected and autonomous vehicles. Expensive bespoke critical communications systems will increasingly migrate to commercial 5G networks, with dedicated network slices enabling specific network characteristics in support of exacting key performance indicators and service levels. Network slicing enables bespoke solutions to be engineered across common network infrastructure with a lower total cost of ownership than today's parallel network approach. Private and semi-private networks will be enabled to provide specific coverage and/or performance characteristics to address exacting requirements of individual vertical industries.

Acknowledgements

The author acknowledges the significant contribution made by his colleagues, Chris Simcoe and Tom Curry of BT Technology, to the development of the network architecture model discussed in this chapter. The author would like to thank Professor Nigel Linge from the University of Salford and Julian Divett from BT Technology for their review of this chapter and constructive feedback which improved the final text.

References

1 ITU-R, Recommendation ITU-R M.2083-0 909/2015) IMT Vision – Framework and overall objectives of the future development of IMT for 2020 *and beyond*, Geneva: ITU-R, 2015.

2 NGMN. *NGMN 5G White Paper*. NGMN Ltd, Frankfurt, 2015.

3 Inmarsat plc. & Deutsche Telekom AG. European Aviation Network. Inmarsat plc. & Deutsche Telekom AG, 2018. https://www.europeanaviationnetwork.com/. [Online: accessed 14 July 2018].

4 NGMN. *5G Extreme Requirements: Radio Access Network Solutions*. NGMN Ltd, Frankfurt, 2018.

5 3GPP. 3GPP. 3GPP, 2018. http://www.3gpp.org/release-15. [Online: Accessed 14 July 2018]

6 3GPP. TR 38.801 Study on new radio access technology: Radio access architecture and interfaces. 3GPP, Sophia Antipolis, 2017.

7 3GPP. 3GPP TS 38.470 V1.0.0 (2017-12) F1 general aspects and principles. 3GPP, Sophia Antipolis, 2017.

8 Metro Ethernet Forum. Carrier Ethernet Specifications. 2018. Available: https://www.mef.net/. [Online: Accessed 20 July 2018]

9 A. Sutton. Mobile backhaul evolution: from GSM to LTE-Advanced, in *Access, Fronthaul and Backhaul Networks for 5G & Beyond*, London, IET, 2017, pages 272–279.

10 Ericsson AB, Huawei Technologies Co. Ltd, NEC Corporation and Nokia. eCPRI Specification V1.0 (2017-08-22) Common Public Radio Interface: eCPRI Interface Specification. Ericsson AB, Huawei Technologies Co. Ltd, NEC Corporation and Nokia, 2017.

11 MulteFire Alliance. MulteFire. 2018. https://www.multefire.org/. [Online: Accessed 15 July 2018].

12 3GPP. 3GPP TR 38.913 V14.3.0 Study on Scenarios and Requirements for Next Generation Access Technologies. 3GPP, Sophia Antipolis, 2017.

13 A. Sutton. 5G Network Architecture, Design and Optimisation. in *IET 5G – State of Play*, London, 2018.

2

Industrial Wireless Sensor Networks and 5G Connected Industries

Mohsin Raza, Sajjad Hussain, Nauman Aslam, Hoa Le-Minh and Huan X. Nguyen

The fourth industrial revolution requires a high degree of automation, increased efficiency, system flexibility, improved reliability, optimized production, system safety, and security for effective optimized and greener industries. To ensure improved performance and to transform large industrial infrastructures into a single optimized unit, the communication interlink between large industrial blocks is of utmost importance. 5G offers flexible technology blocks to ensure desirable outcomes by implementing Ultra-Reliable and Low-Latency Communications (URLLC) within the industries to offer improved process control and automation. In this chapter, a detailed discussion of industrial revolutions, technology improvements, industrial processes and control requirements, and communication deadlines is provided. The chapter also discusses novel schemes to optimize communications within industrial processes to ensure desired time deadlines and reliability constraints. The chapter also summarizes the role of 5G technologies in ensuring the future of modern day industries.

2.1 Overview

Industries play an important role in the global economy. The significance of industries and growing competitions drive the technological development of these sectors, which results in continuous improvement and potential research in the areas. These developments enable consistent advancement in industries and closely related fields [1, 2]. With the introduction of automation and process control, the production capabilities of industries have been significantly increased and thus resulted in further growth in industries [3]. The transformation of industries from non-automated to fully automated intelligent systems increases the system complexity, which requires rapid changes in the existing technologies to support complex operations.

The involvement of automation, process control and system intelligence in industries has also altered the communication systems requirements. The changes in the industrial systems require more sophisticated real-time and reliable communication infrastructure.

Industrial Wireless Sensor Networks (IWSNs) have been widely investigated to support the changing industries. IWSNs offer several advantages to support the dynamic

Enabling 5G Communication Systems to Support Vertical Industries, First Edition.
Edited by Muhammad Ali Imran, Yusuf Abdulrahman Sambo and Qammer H. Abbasi.
© 2019 John Wiley & Sons Ltd. Published 2019 by John Wiley & Sons Ltd.

and complex operations in industries. With easy deployment, unrestricted mobility and intelligent information routing abilities, IWSNs emerge as a potential communications solution for industries. Besides, IWSNs are a low-cost solution compared to wired solutions. There are several advantages of using IWSNs, though it also poses limitations in some aspects of communications in industrial applications.

Over the years IWSNs have been thoroughly investigated. The keen interest of the research community and industry provided notable improvements in the technology. In the past few years, many IEEE and industrial standards were introduced. Some of the well-known IEEE standards to support communications within the industries include IEEE 802.15.4 and IEEE 802.15.4e [4, 5]. These standards provide Physical and MAC layer specifications for low data rate communications of sensor networks. Using the Physical and MAC layer specifications of IEEE standards, many industrial protocols were also developed which used customized features of IEEE standards along with the specification of upper layers of the OSI model. Some notable industrial standards widely used in industries at present include Zigbee, WirelessHART, ISA100.11a, 6LoWPAN Wia-PA and OCARI [6–10]. All these developments resulted in dedicated research in IWSNs and possible communication solutions to fulfil the ever-changing requirements of industrial processes.

The wider applicability of sensor networks in industries resulted in rapid growth in the field, while in the past few years several improvements were suggested in the MAC and Network layer. These contributions pave the path for future deterministic, low-latency, reliable, self-healing sophisticated networks with URLLC. The URLLC in 5G networks offers a potential technology to minimize the operability issues in the existing IWSNs and legacy systems with the assurance of low-latency and reliable communications.

2.2 Industrial Wireless Sensor Networks

2.2.1 Wired and Wireless Networks in Industrial Environment

The wireless technology provides a potential solution for improved communications in an industrial process. However, the high interference in industries also makes the desirable communications reliability very challenging. Therefore, to offer the desired reliability wired communications were widely accepted in the industries in the last decade. Some of the well-known and commonly implemented wired communications solutions include Ethernet/IP, Real-Time Ethernet (RTE), High-Speed Ethernet (HSE), CC-LINK IE, EtherCAT, Sercos III, CAN, H1, Profibus DP and HART. As a trade-off for high reliability, the usual sacrifices are low flexibility, hectic deployment, debugging and expensive link costs.

The failure in the provision of flexible, low-cost intelligent networks using wired solutions proved resourceful in development of alternate solutions. These limitations also inspired the formation of dedicated research groups and industrial liaison to investigate alternatives, leading to the development of several IEEE and industrial standards.

2.2.2 Transformation of WSNs for Industrial Applications

Traditional WSNs consist of spatially distributed autonomous devices used to sample the environmental parameters for low reliability and time insensitive monitoring applications [11, 12]. Due to the distributed nature, small size and short range of communications, wireless sensor motes are well suited for forming dense sensing networks [13]. The on-demand network formation in WSNs allows incorporation of new nodes and network updates on the run, which is suitable for dynamic network formation.

The suitability of WSNs for industrial applications due to its unique characteristics allowed its implementation within the industries. However, the basic characteristics of traditional WSNs were unsuitable for critical industrial applications. Therefore a steady change in WSNs is seen to meet the requirements of highly critical processes and real-time decision-making [14]. The communications in highly sensitive plants and industries introduce several challenges. A well-developed communications infrastructure is required to meet strict reliability constraints and time deadlines within the industries.

The stability of underlying control processes within the industries mainly depends on the communications link. Therefore, to run processes effectively, it is required to communicate sensed values at regular intervals within bounded delays. Apart from the latency bounds, reliability and energy efficiency are also important aspects of wireless networks, which ensure deterministic network attributes and extended network life. The network lifetime while mainly depends on the application at hand, the significance of effective transmission power control strategies, sleep scheduling, energy-efficient data routing and communication scheduling cannot be undermined. Along with the advances in traditional WSNs, 5G assisted technologies also offer solutions to overcome communications challenges in IWSNs [15].

2.2.3 IWSN Architecture

IWSN architecture offers various attributes to suit a variety of industrial applications. These attributes individually as well as collectively specify the suitability of resultant IWSN for certain set of applications. Some of these attributes include hardware specifications, network architecture/topology, multiple access schemes, data assimilation mechanisms, information security and interoperability. Various combinations of different attributes define whether the developed IWSN architecture is suitable for critical or non-critical applications.

IWSNs are formed of a large number of nodes. Each node consists of radio, processor, Input/Output (I/O) ports, memory, sensors and battery. In some cases, nodes may also have an energy harvesting mechanism or multiple radios to facilitate different applications [16, 17]. These deviations from traditional node specifications offer added benefits including network lifetime optimization, duplex communications etc. [18, 19].

The suitability of IWSN for certain application greatly depends on its architecture. A particular IWSN architecture (node specifications, network topology, communications frequency etc) offers certain characteristics and with the change in the base architecture, these characteristics can change.

Considering one aspect out of many, changes in network topology can have diverse effects on the performance of IWSNs. Nodes in the network can be connected as a star topology, mesh topology or tree topology to reap different benefits. In star topology, nodes are directly connected to the gateway, which can minimize delay in communications. However, when the connected nodes increase in the cluster the reliability is severely affected. On the other hand, in the tree topology, while a larger number of nodes can be handled by single gateway, delay is notably increased.

The channel access schemes also play an import role in defining the attributes of IWSN. Time Division Multiple Access (TDMA) scheme limits the communication time of an individual node with the gateway whereas Carrier Sense Multiple Access/Collision Avoidance (CSMA/CA) suffers reliability with the increase in connected nodes. Apart from these, other topologies have their own pros and cons.

The network architecture is also one of the important attributes of the IWSN. The selection of network architecture mainly depends on the application at hand. The flat architecture offers low complexity and works well in small networks but with the expansion of the network, performance on the network is severely affected. In addition, the routing tables become encumbered and some nodes are overloaded, which affects network lifetime and results in congestion development. Alternatively, hierarchical architecture offers quick access to the cluster heads and gateway and is more suitable for heterogeneous networks. However, a hierarchical architecture is more complex and cluster formation and suitable affiliation establishment add computational load. A graphical representation of different aspects of IWSN architecture is presented in Figure 2.1.

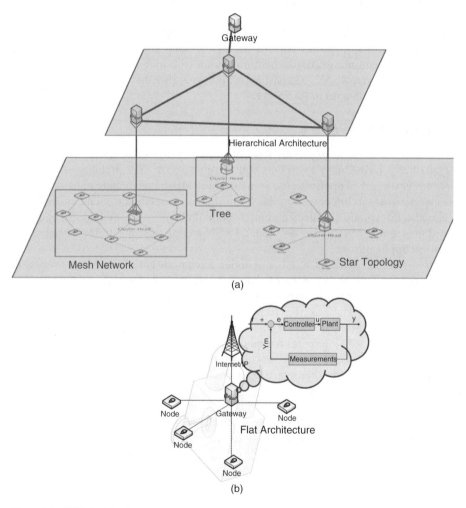

Figure 2.1 IWSN Architecture.

Table 2.1 Traffic categories and affiliated attributes in IWSNs.

Sr.	Traffic Category	Priority	Applications	Tolerance			Medium Access Control
				Time Constraint	Reliability		
1	Safety/Emergency traffic	Very high	Emergency/Alarms	Few milliseconds	High-reliability requirements		Prioritized slotted access
2	Critical control traffic	High	Closed loop process control/critical feedback	Tens of milliseconds	High reliability		TDMA/prioritized CSMA
3	Low-risk traffic	Medium	Slow process control/pow-priority interrupts	Seconds to hours	Medium reliability with occasional packet misses acceptable		CSMA/slotted access with high-priority overwrite ability
4	Monitoring traffic	Low	Monitoring application/static feedback	minutes to hours	Low-reliability requirements		Low importance data Retransmissions are avoided

2.3   Industrial Traffic Types and Its Critical Nature

Numerous researches [20–22] divide communications traffic in industries into various classes. This classification depends on the underlying process, equipment and type of operation performed. In this chapter, the traffic in industries is categorized into four groups. Table 2.1 lists these traffic types and their main attributes.

2.3.1   Safety/Emergency Traffic

In industries, safety/emergency communications are very important and have the highest priority. This communication ensures to minimize the impact on humans and industrial equipment. This type of traffic ensures the safekeeping and smooth running of the plant. Due to its importance, failure in this type of traffic inflicts dangerous consequences. Due to this nature it is prioritized over all other communications in the industry in wired Fieldbus protocols. The same is expected in IWSNs, where systems are carefully modelled to offer high reliability and low latency.

2.3.2   Critical Control Traffic

Control information has a relatively higher significance compared to slow process control and monitoring traffic. The timely and reliable communication of critical control traffic ensures the smooth running of the processes and restricts violation of critical thresholds to the minimum. The critical control information is usually associated with regulatory control systems and mainly targets the most critical processes in the industry. Any negligence in the critical control information may result in the emergency trigger. For most cases, critical control traffic is modelled as synchronous information load, which occupies constant bandwidth.

2.3.3   Low-Risk Control Traffic

Low-risk control traffic mainly reports the low-risk data to the control centre and occasional actions are expected in response. This type of control traffic is relatively less time-critical in nature than the safety and critical control traffic and occasional misses in the delivery of this information would not trigger any critical hazards.

2.3.4   Periodic Monitoring Traffic

Monitoring traffic is low reliability and high latency tolerant traffic, which is not directly used in any actuators in the process control industry. Rather the information is mostly accumulated for analysis in the later stages to support future upgrading plans.

Apart from the critical nature of the communication, the traffic originating sources are also categorized into multiple groups based on the pattern of their channel access and hence the communication originating from them. Traffic with respect to channel access mechanisms is categorized into four significant classes: asynchronous, synchronous, slotted and multi-channel [22–25].

2.3.5   Critical Nature and Time Deadlines

Conventional WSNs usually target less critical applications with low urgency requirements where delay in the sensed information and failure in quick response does not trigger alarms. Unfortunately, in industrial applications, it is not the case. In industries, therefore, it is of utmost importance how the IWSNs respond to the information generated in a plant, especially whose which are more critical in nature.

The industrial processes and control systems are relatively time-constrained and highly critical processes with strict bounds on the communication reliability and response time. The link between the processes and the control unit plays a very crucial role in ensuring the time deadlines. Therefore, the reliability of the link and the timely communication need to be prioritized. Furthermore, some processes are more critical than others and therefore need prioritized access to the channel compared to others.

To ensure the smooth running of the processes and to build tolerance against abnormalities, a cross-layer design must be devised. It is also important to meet the deadlines set forth by the International Society of Automation. A few requirements are listed in Table 2.2.

Table 2.2 Typical delay and update requirements for industrial processes [26–28].

Applications	Battery Life	Security Requirements	Update Frequency
Monitoring and Supervision			
Vibration sensor	3 yrs	Low	sec – days
Pressure sensor	3 yrs	Low	1 sec
Temperature sensor	3 yrs	Low	5 sec
Gas detection sensor	3 yrs	Low	1 sec
Others/Data acquisition	3 yrs	Low	> 100 ms
Closed Loop Control			
Pressure sensor	5 yrs	Medium	Up to 500 ms
Temperature sensor	5 yrs	Medium	Up to 500 ms
Flow sensor	5 yrs	Medium	Up to 500 ms
Torque sensor	5 yrs	Medium	Up to 500 ms
Variable speed drive	5 yrs	Medium	Up to 500 ms
Interlocking and Control			
Proximity sensor	5 yrs	High	Up to 250 ms
Motor	5 yrs	High	Up to 250 ms
Valve	5 yrs	High	Up to 250 ms
Protection relays	5 yrs	High	Up to 250 ms
Machinery and tools	3 yrs	High	10 ms
Motion control	3 yrs	High	1 ms

2.4 Existing Works and Standards

The communications in industries can be categorized either as wired or wireless. In the last decade, wired networks dominated the industrial communications, whereas wireless networks, though slowly progressed in last decade, have nevertheless received much fame due to research developments and advancements in wireless industrial solutions. In the past few years, developments in microelectromechanical systems (MEMS) technology have resulted in low-cost motes with improved processing, memory and efficiency. Further developments in software and communications protocols also provided better and optimized communications and data sampling abilities. The developments on all aspect of sensor networks, especially the industrial standards and protocols, assisted in the large-scale implementation of WSNs in industrial applications. Some of the communications standards are discussed in detail where recent improvements are presented.

2.4.1 Wireless Technologies

Wireless communication standards addressing communication issues in WPAN are standardized by 802.15, a working group of IEEE 802.15 consists of ten groups, ranging from Bluetooth to visible light communication. Communications in industries and standards in IWSNs and covered in IEEE 802.15.4. Since there are some other technologies that have certain significance in industrial applications, therefore a brief review of these technologies is also included.

Various communication standards have been adopted in the industries for different applications. Some of the well-known technologies and standards include Wi-Fi, Bluetooth, Zigbee and Ultra-Wide Band (UWB).

IEEE 802.11 standard, also referred to as Wi-Fi, operates in ISM band (2.4GHz). It provides good data rates for close range sensing applications [29]. It also allows networks with a large number of nodes. With high data rates and scalable network capabilities, Wi-Fi modules consume more energy. They also cost more than the typical wireless sensor motes. High energy consumption affects network lifetime and results in further maintenance costs.

Bluetooth, which is based on IEEE 802.15.1, provides an energy-efficient solution for communications. For communication purposes, 79 channels, each of 1MHz bandwidth, are used [30]. The Bluetooth devices are relatively low cost and offer improved security. Use of Bluetooth offers security and cost-efficiency. However, it limits the total number of connected nodes. Earlier Bluetooth standards only facilitated eight nodes. This restriction was eliminated in Bluetooth 4.0 and later standards but connected devices work in slave-master interconnection and lack flexibility for multi-hop communication, which limits the scope of industrial applications where Bluetooth can fit in.

UWB is usually considered a solution for industrial applications with relatively short communication distances. It offers high data rates, transmitted in very short pulses. It uses 15 channels for communications and offers high data rates in the range of 110Mbps. However, the energy requirements are relatively high for battery-operated devices.

Zigbee is an industrial protocol which uses the Physical and MAC layer specifications of IEEE 802.15.4 for low data rate applications. With 16 channels, 2 MHz bandwidth and 16-bit address, it can support networks sizes of up to 65000 approximately. The peer-to-peer communications facilitates mesh networking and multi-hop communications. Since Zigbee is defined for battery-operated devices, it therefore targets long network lifetime. The CSMA/CA scheme is implemented in it to avoid collision and channel congestion.

2.4.2 Industry-Related IEEE Standards

Some of the well-known wireless industrial protocols use Physical and MAC layer specifications of IEEE standards for WPAN to provide a baseline. Two main IEEE standards, IEEE 802.15.4 and IEEE 802.15.4e, are used in Zigbee, WirelessHART, ISA100.11a, 6LoWPAN and MiWi. Details of these IEEE standards are as follows.

2.4.2.1 IEEE 802.15.4

IEEE 802.15.4 specifies the Physical and MAC layer for low-power sensor networks. The standard specifies Physical and MAC layer properties. In these standards Physical layer cover details related to modulation schemes, RF parameters, frequency requirements, transmission power, channel details, spreading parameters and channel assignment. It also specifies the recommended receiver sensitivity and link quality measure along with the channel state assessment. Table 2.3 presents some of the Physical layer parameters.

The MAC layer describes terms to access the physical channel, which includes beacon generation, frame synchronization, an affiliation of sensor nodes to PAN, medium access strategies (CSMA/CA for congestion avoidance), security and reliable link establishment between the MAC entities [31].

Table 2.3 PHY layer parameters for IEEE 802.15.4 [31].

		Spreading Parameters		Data Parameters		
PHY (MHz)	Frequency Band (MHz)	Chip rate (kchip/s)	Modulation	Bit Rate (Kb/s)	Symbol Rate (Ksymbols/s)	Symbols
780	779–787	1000	O-QPSK	250	62.5	16-ary orthogonal
780	779–787	1000	MPSK	250	62.5	16-ary orthogonal
868/915	868–868.6	300	BPSK	20	20	Binary
	902–928	600	BPSK	40	40	Binary
868/915 (optional)	868–868.6	400	ASK	250	12.5	20-bit PSSS
	902–928	1600	ASK	250	50	5-bit PSSS
868/915 (optional)	868–868.6	400	O-QPSK	100	25	16-ary orthogonal
	902–928	1000	O-QPSK	250	62.5	16-ary orthogonal
950	950–956	—	GFSK	100	100	Binary
950	950–956	300	BPSK	20	20	Binary
2450 DSSS	2400–2483.5	2000	O-QPSK	250	62.5	16-ary orthogonal

2.4.2.2 IEEE 802.15.4e

Unlike IEEE 802.15.4, IEEE 802.15.4e is particularly defined to facilitate communications in critical industrial applications. It uses TDMA-based channel access to ensure higher reliability. The use of TDMA also ensures guaranteed channel access and eliminates uncertainty present in contention-based schemes. Since IEEE 802.15.4e is optimized for critical operation within industries, therefore a specific framework for Low-Latency Deterministic Networks (LLDN), IEEE 802.15.4e LLDN, is defined to facilitate communications for time-constrained and reliability-sensitive communications. To optimize the retransmissions of failed communications, shared slots are introduced which use CSMA/CA-based channel access.

Other Significant Research Contributions In the past few years, several wireless communication solutions have also been proposed by the research community. These solutions targeted improvements in different aspects of wireless communications within the industries. Due to the extensive automation introduced within the industries, the developments in IWSNs targeted improved reliability, latency, network lifetime and network formation. Most of the proposed researches targeted MAC layer and network layer optimization. Given the importance of both the layers, the improvements in their functionality enabled IWSNs to offer desired Quality of Service (QoS) for critical industrial process and control systems.

In the past few years, a notable contribution can be seen in the network layer optimizations. Network layer plays a very important role in the real-time and reliable delivery of information to the control centre in IWSNs. To meet the critical needs of the industrial applications, many protocols are proposed. A flooding-based routing scheme is presented in [32]. In this paper, the authors proposed a reliable and real-time algorithm, which reduces the overhead of routing tables and exploits multipath diversity to improve both reliability and real-time data delivery. Simulations are used to ensure performance improvements in reliability, network recovery time and latency. In [33], an efficient route selection algorithm, InRout, is presented. It takes into account the local information and implements Q-learning techniques to optimize route decision and enhance reliability and real-time data delivery. The authors claim an improvement of up to 60% in packet delivery compared to existing schemes and also cuts down the control overhead. The authors in [34] proposed a two-hop routing protocol targeting reliability and real-time performance enhancement. The paper improves real-time data delivery but lacks in reliable communication. Apart from these, many other routing protocols were proposed for monitoring applications in industrial environments. Due to less critical nature of applications in these protocols, focus on reliability and real-time data delivery are not the primary concerns. An exhaustive list of such protocols can be found in [35–37].

The scope of MAC protocols has significantly changed over the years. Earlier on, the research in WSNs targeted extended network lifetime and longer operation of remote sensor networks. A more detailed review of such MAC protocols can be found in [38, 39]. Moreover, a thorough survey of energy-efficient MAC protocols can be found in [40]. However, in recent years a trade-off has been established in ensuring longer network operation along with the desired quality of service. Currently, MAC protocols can never target energy efficiency as only design concern as much as they cannot rely on best effort data delivery [41].

In the past few years, a large number of MAC protocols were introduced. These protocols can be categorized in a number of ways. An extensive list of such protocols can be found in [42–45]. In [43], authors classify MAC protocols in four categories: random, periodic, slotted and hybrid. Another classification can be found in [46], where MAC protocols are classified as asynchronous, synchronous, slotted and multi-channel. Each of these categories have their unique benefits. Asynchronous protocols occasionally communicate and run on low duty cycle, which enables long network lifetime. However, these protocols fail to offer efficient communication within the network. Asynchronous MAC protocols are thoroughly reviewed in [25]. The limitations of asynchronous MAC protocols are addressed in synchronous protocols. However, due to the regular communication requirements, congestion appears to be a major challenge. Slotted channel access schemes using TDMA resolve the congestion issue. However, channel usage is relatively low. Apart from that the bandwidth and capacity are also limited in such schemes.

Multi-channel schemes benefit from both TDMA and Frequency Division Multiple Access (FDMA), which improve channel usage. Some other schemes also target MAC layer optimization using both TDMA and CSMA/CA.

2.5 Ultra-Reliable Low-Latency Communications (URLLC) in IWSNS

Despite the frequent developments and added benefits of IWSNs, researchers are still struggling to offer substantial solutions for the improved reliability and real-time data delivery in IWSNs to match the strict deadlines as needed for the closed loop process control and URLLC [47, 48]. URLLC is mandatory for IWSNs, especially when dealing with the emergency communications and regulatory control feedback systems. To improve the overall acceptability of IWSNs, and to cope with fast-paced improvements in industry and protocol stack developments, restructuring and procedural changes for URLLC in IWSNs are very important. Furthermore, the inclusion of Machine-to-Machine (M2M) communications in 5G, offers potentially benefitting framework, targeting three main aspects of IWSNs: 1) support for a large number of low-rate network devices, 2) sustaining a minimal data rate in all circumstances to satisfy the feedback control requirements and 3) enabling very low-latency data transfer [49]. Addressing such requirements in IWSNs using 5G M2M infrastructure needs further architectural improvements, and procedure restructuring is necessary for 5G M2M infrastructure to address such requirements in IWSNs.

In this section, a hybrid multi-channel scheme for performance and throughput enhancement of IWSNs is presented. The scheme utilizes multiple frequency channels to increase the overall throughput of the system along with the increase in reliability. Use of multiple channels allows URLLC in IWSNs. A special-purpose frequency channel is also defined, which facilitates the failed communications by retransmissions where the retransmission slots are allocated according to the priority level of failed communications of different nodes. A scheduler is used to formulate priority-based scheduling for retransmission in TDMA-based communication slots of this channel. Furthermore, in carrier-sense multiple access with collision avoidance (CSMA/CA)-based slots, frequency polling is introduced to limit the collisions. Mathematical modelling for performance metrics is also presented. The performance of the scheme alongside IEEE 802.15.4e is also presented.

The scheme considers the impact of multiple channels compared to single-channel schemes and evaluates improvements in a number of nodes per cluster, communication reliability and overall throughput. To evaluate the performance of the proposed scheme, a mathematical formulation of the possible scenarios for typical IWSNs as well as for the proposed scheme in IWSNs is presented as follows.

The communication from all the affiliated nodes in a cluster periodically originates in a specified time slot of every superframe. The success of each individual communication is dependent on the channel conditions and the probability of success of an individual communication, which is represented with p, whereas the total successes in every time slot are modelled as binomial (p, w) distribution.

In a typical IEEE 802.15.4e system, using a single frequency channel, the frame error rate depends on several factors including the number of high-priority nodes, multipath fading, dispersion, reflection, refraction, interference, distance, congestion, transmission power restrictions and receiver sensitivity. To demonstrate the frame error rate for various high-priority nodes, a mathematical formulation is presented in (1), where k is the number of high-priority nodes.

$$P(frame\_error\_rate|H = 1) = \sum_{x=1}^{k} \binom{k}{x} (1 - p)^x p^{k-x} \tag{2.1}$$

The use of multiple channels introduces a significant improvement in the throughput of the system but the reliability in such cases is dependent on the number of RC channels and ratio of RC and SP channels. A mathematical expression for the frame error rate in case of multiple channel scenarios is given in (2) to (4). The maximum number of high-priority nodes are limited to k per RC channel whose maximum numbers are limited to H. Based on the total number of the high-priority nodes permitted (k) and total communications in a single frame (n), limits can be generalized to w_1 and w_2 for (2) to (4), where $w = k \times H$, $w_1 = n/2$ and $w_2 = n$.

$P(frame\_error\_rate|(H > 1)\&(w \leq w_1))$

$$= \frac{\left[\left(\sum_{x=1}^{w} \left(\binom{w}{x} (1 - p)^x p^{(w-x)} \times \left(\sum_{y=1}^{x} \binom{x}{y} (1 - p)^y p^{(x-y)} \right)^2 \right) \right) \right]}{H}, \tag{2.2}$$

$P(frame\_error\_rate|(H > 1)\&(w_1 < w \leq w_2))$

$$= \frac{\left[\begin{array}{l} \left(\sum_{x=1}^{w_1} \left(\binom{w}{x} (1 - p)^x p^{w-x} \times \left(\sum_{y=1}^{x} \binom{x}{y} (1 - p)^y p^{x-y} \right)^2 \right) \right) \\ + \left\{ \sum_{x=w_1+1}^{w} \left(\binom{w}{x} (1 - p)^x p^{w-x} \right. \right. \\ \left. \left. \times \left\{ \left(\sum_{y=1}^{z=n-x} \binom{x}{y} (1 - p)^y p^{x-y} \right)^2 + \left(\sum_{v=z+1}^{x} \binom{x}{v} (1 - p)^v p^{x-v} \right) \right\} \right) \right\} \end{array} \right]}{H}$$

$$\tag{2.3}$$

$P(frame\_error\_rate|(H > 1)\&(w > w_2)$

$$= \frac{\left[\begin{array}{l}\left\{\sum_{x=1}^{w_1}\left(\binom{w}{x}(1-p)^x p^{w-x} \times \left(\sum_{y=1}^{x}\binom{x}{y}(1-p)^y p^{x-y}\right)^2\right)\right\} \\ + \left\{\sum_{x=w_1+1}^{w_2}\left(\binom{w}{x}(1-p)^x p^{w-x}\right.\right. \\ \qquad \left.\left.\times \left\{\left(\sum_{y=1}^{z=n-x}\binom{x}{y}(1-p)^y p^{x-y}\right)^2 + \left(\sum_{v=z+1}^{x}\binom{x}{v}(1-p)^v p^{x-v}\right)\right\}\right)\right\} \\ + \left\{\sum_{x=w_2+1}^{w}\left(\binom{w}{x}(1-p)^x p^{w-x}\right)\right\}\end{array}\right]}{H}$$

(2.4)

The performance of the multiple channel scheme with a dedicated retransmission channel (SP channel) is evaluated as a function of the probability of successful communication of an individual node, total number of parallel data streams, i.e. the number of communication channels (RC channels) and the number of high-priority nodes (k) trying to communicate in a single superframe duration. Twenty transmissions in a single superframe are used ($n = 20$), so w_1 and w_2 are set to 10 and 20 respectively.

For communication, a superframe duration of 10 ms is used. The performance is evaluated, based on the number of high-priority nodes (k) communicating within the superframe where the total number of nodes trying to attempt a communication in a single superframe is limited to twenty. The frame error rate is evaluated for different channel conditions under which the probability of communication failure $(1 - p)$ is represented by q.

By introducing an SP channel in the typical IEEE 802.15.4e system as expressed in the presented scheme, a notable error rate reduction (with error dropping from 5% in IEEE 802.15.4e to 0.001% in proposed scheme with two RC channels for $q = 0.001$ and *High priority nodes* $= 5$, see Figures 2.2 and 2.3) can be seen in the communication of

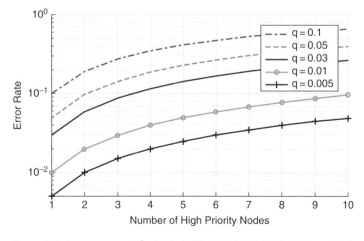

Figure 2.2 Frame error rate for typical IEEE 802.15.4e (LLDN) with one RC channel and no SP channel.

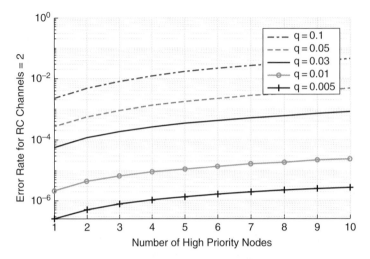

Figure 2.3 Frame error rate for the proposed scheme with two RC channel and one SP channel.

high-priority nodes. Since the rest of the n-k nodes communicating in the network are considered as low priority nodes, so the failure in the communication of these nodes is not critical and will not affect the performance of feedback control systems. With the retransmission of failed communication in RC channel through the SP channel, a notable improvement in the reliability of communication can be seen (with error rate dropping below 3% in all investigated cases for two RC channels). A similar conclusion can be deduced by comparing the error rate of IEEE 802.15.4e presented in Figure 2.2 and proposed multi-channel scheme with two RC channel and one SP channel, presented in Figure 2.3. To further evaluate the effect of using multi-channel scheme for a higher number of parallel data streams, the error rate is evaluated for the cases with one SP channel as well as 6 and 9 RC channels, as presented in Figures 2.4 and 2.5 respectively.

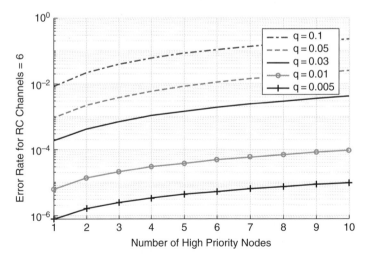

Figure 2.4 Frame error rate for the proposed scheme with six RC channels and one SP channel.

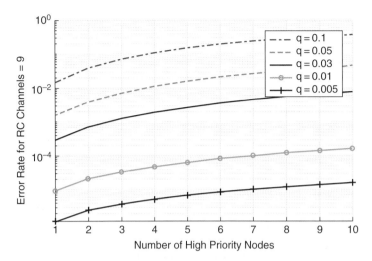

Figure 2.5 Frame error rate for the proposed scheme with nine RC channels and one SP channel.

2.6 Summary

The fourth industrial revolution redefines robotics, automation, process control, system optimization, machine intelligence and distributed control systems. The use of effective communications is mandatory to maintain the operations of present-day industries with the utmost accuracy. It is also important that different classes of traffic are appropriately handled. URLLC offers one such framework to support communications requirements within the industry. The work presented in this chapter uses both frequency division and time division along with CSMA/CA-based channel access with failed communications scheduling to improve communications within diverse industrial wireless networks.

References

1 C. Ying-Chyi, L. Ching-hua, and C. Pao-Long. Using theory of constraints to find the problem about high level inventory in the aerospace industry. In *Technology Management for Global Economic Growth (PICMET), 2010 Proceedings of PICMET'10*, pages 1–10, 2010.

2 M. Raza, N. Aslam, H. Le-Minh, S. Hussain. A novel MAC proposal for critical and emergency communication in industrial wireless sensor networks. *Computers and Electrical Engineering*, 72:976–989, 2018.

3 A. Z. Xu. (2014, 2020): Future automation. http://www.controleng.com/single-article/2020-future-automation/c33ed4679973dcb1ed2f53411520088d.html.

4 IEEE Standard for Local and Metropolitan Area Networks – Part 15.4: Low-Rate Wireless Personal Area Networks (LR-WPANs) Amendment 1: MAC sublayer. *IEEE Std 802.15.4e-2012 (Amendment to IEEE Std 802.15.4-2011)*, pages 1–225, 2012.

5 IEEE Standard for Information technology – Local and metropolitan area networks – Specific requirements – Part 15.4: Wireless Medium Access Control (MAC) and Physical Layer (PHY) Specifications for Low Rate Wireless Personal Area

Networks (WPANs). *IEEE Std 802.15.4-2006 (Revision of IEEE Std 802.15.4-2003),* pages 1–320, 2006.

6 K. Al Agha, M.-H. Bertin, T. Dang, A. Guitton, P. Minet, T. Val, et al. Which wireless technology for industrial wireless sensor networks? The development of OCARI technology. *Industrial Electronics, IEEE Transactions on,* vol. 56, pages 4266–4278, 2009.

7 H. C. Foundation. (2016). *WirelessHART Overview.* Available: http://en.hartcomm .org/hcp/tech/wihart/wireless_overview.html.

8 (2016). *6loWPAN Active Drafts.* Available: https://datatracker.ietf.org/doc/search/? name=6loWPAN&activeDrafts=on&rfcs=on.

9 (2016). *Utility Industry | The Zigbee Alliance.* Available: http://www.zigbee.org/ what-is-zigbee/utility-industry/.

10 (2016). ISA-100 Wireless Compliance Institute – ISA-100 Wireless Compliance Institute – Official Site of ISA100 Wireless Standard – isa100wci.org. Available: http:// www.isa100wci.org/.

11 T. Instruments, *What Is a Wireless Sensor Network (White Paper),* 2012.

12 M. Raza, G. Ahmed, N. M. Khan, M. Awais and Q. Badar. A comparative analysis of energy-aware routing protocols in wireless sensor networks. *2011 International Conference on Information and Communication Technologies,* Karachi, pages 1–5, 2011. doi: 10.1109/ICICT.2011.5983546

13 I. F. Akyildiz and M. C. Vuran, *Wireless Sensor Networks,* vol. *4: John Wiley & Sons,* 2010.

14 Q. Pham Tran Anh and K. Dong-Sung. Enhancing real-time delivery of gradient routing for industrial wireless sensor networks. *Industrial Informatics, IEEE Transactions on,* vol. 8, pages 61–68, 2012.

15 V. Ç. Güngör and G. P. Hancke, *Industrial Wireless Sensor Networks: Applications, Protocols, and Standards*: CRC Press, 2013.

16 M. J. Miller and N. H. Vaidya. A MAC protocol to reduce sensor network energy consumption using a wakeup radio. *Mobile Computing, IEEE Transactions on,* vol. 4, pages 228–242, 2005.

17 M. Raza, N. Aslam, H. Le-Minh, S. Hussain, N. M. Khan, Y. Cao. A Critical Analysis of Research Potential, Challenges and Future Directives in Industrial Wireless Sensor Networks. *IEEE Communications Surveys & Tutorials,* 2017.

18 J. Shouling, L. Yingshu, and J. Xiaohua. Capacity of dual-radio multi-channel wireless sensor networks for continuous data collection. In *INFOCOM,* 2011 *Proceedings IEEE, pages* 1062–1070, 2011.

19 S. Sudevalayam and P. Kulkarni. Energy Harvesting Sensor Nodes: Survey and Implications. *Communications Surveys & Tutorials, IEEE,* vol. 13, pages 443–461, 2011.

20 Z. Meng, L. Junru, L. Wei, and Y. Haibin. A priority-aware frequency domain polling MAC protocol for OFDMA-based networks in cyber-physical systems. *Automatica Sinica, IEEE/CAA Journal of,* vol. 2, pages 412–421, 2015.

21 S. Wei, Z. Tingting, F. Barac, and M. Gidlund. PriorityMAC: A Priority-Enhanced MAC Protocol for Critical Traffic in Industrial Wireless Sensor and Actuator Networks. *Industrial Informatics, IEEE Transactions on,* vol. 10, pages 824–835, 2014.

22 T. Zheng, M. Gidlund, and J. Akerberg. WirArb: A New MAC Protocol for Time Critical Industrial Wireless Sensor Network Applications. *Sensors Journal, IEEE,* 16(7):1-1, 2015.

23 W. Dong, J. Yu, J. Wang, X. Zhang, Y. Gao, C. Chen, et al. Accurate and robust time reconstruction for deployed sensor networks. *Networking, IEEE/ACM Transactions on*, 24(4):2372–2385, 2015.

24 I. H. Hou. Packet scheduling for real-time surveillance in multihop wireless sensor networks with lossy channels. *Wireless Communications, IEEE Transactions on*, vol. 14, pages 1071–1079, 2015.

25 M. Doudou, D. Djenouri, and N. Badache. Survey on latency issues of asynchronous MAC protocols in delay-sensitive wireless sensor networks. *Communications Surveys & Tutorials, IEEE*, vol. 15, pages 528–550, 2013.

26 P. Neumann. Communication in industrial automation—What is going on? *Control Engineering Practice*, vol. 15, pages 1332–1347, 2007.

27 D. Christin, P. S. Mogre, and M. Hollick. Survey on wireless sensor network technologies for industrial automation: The security and quality of service perspectives. *Future Internet*, vol. 2, pages 96–125, 2010.

28 J. Xia, C. Zhang, R. Bai, and L. Xue. Real-time and reliability analysis of time-triggered CAN-bus. *Chinese Journal of Aeronautics*, vol. 26, pages 171–178, 2013.

29 A. Raheem. Use Of open-source platforms in designing of low-cost prototype wireless sensor mote. MS Electronic Engineering, Electronic Engineering, Mohammad Ali jinnah University, Islamabad, 2013.

30 IEEE Standard for Information technology – Local and metropolitan area networks – Specific requirements – Part 15.1a: Wireless Medium Access Control (MAC) and Physical Layer (PHY) specifications for Wireless Personal Area Networks (WPAN). *IEEE Std 802.15.1-2005 (Revision of IEEE Std 802.15.1-2002)*, pages 1–700, 2005.

31 IEEE Standard for Local and Metropolitan Area Networks – Part 15.4: Low-Rate Wireless Personal Area Networks (LR-WPANs). *IEEE Std 802.15.4-2011 (Revision of IEEE Std 802.15.4-2006)*, pages 1–314, 2011.

32 K. Yu, Z. Pang, M. Gidlund, J. Åkerberg, and M. Björkman. REALFLOW: Reliable real-time flooding-based routing protocol for industrial wireless sensor networks. *International Journal of Distributed Sensor Networks*, vol. 2014, p. 17, 2014.

33 B. Carballido Villaverde, S. Rea, and D. Pesch. InRout – A QoS aware route selection algorithm for industrial wireless sensor networks. *Ad Hoc Networks*, 10(3):458–478, 2012.

34 P. T. A. Quang and D.-S. Kim. Enhancing real-time delivery of gradient routing for industrial wireless sensor networks. *Industrial Informatics, IEEE Transactions on*, vol. 8, pages 61–68, 2012.

35 D. Goyal and M. R. Tripathy. Routing protocols in wireless sensor networks: a survey. In *Advanced Computing & Communication Technologies (ACCT), 2012 Second International Conference on*, 2012, pages 474–480.

36 J. Wan, D. Yuan, and X. Xu. A review of routing protocols in wireless sensor networks. In *Wireless Communications, Networking and Mobile Computing, 2008. WiCOM'08. 4th International Conference on*, 2008, pages 1–4.

37 P. Kumar, M. Singh, and U. Triar. A review of routing protocols in wireless sensor network. *International Journal of Engineering Research & Technology (IJERT)*, vol. 1, 2012.

38 P. Naik and K. M. Sivalingam. A survey of MAC protocols for sensor networks. In *Wireless Sensor Networks*, ed: Springer, 2004, pages 93–107.

39 I. Demirkol, C. Ersoy, and F. Alagoz MAC protocols for wireless sensor networks: a survey. *IEEE Communications Magazine*, 44(4):115–121, 2006.

40 A. Roy and N. Sarma. Energy saving in MAC layer of wireless sensor networks: a survey. In *National Workshop in Design and Analysis of Algorithm (NWDAA), Tezpur University*, India, 2010.

41 P. Suriyachai, U. Roedig, and A. Scott. A survey of MAC protocols for mission-critical applications in wireless sensor networks. *Communications Surveys & Tutorials, IEEE*, 14(2):240–264, 2012.

42 K. Kredo and P. Mohapatra. Medium access control in wireless sensor networks. *Computer Networks*, 51(4):961–994, 2007.

43 K. Langendoen. Medium access control in wireless sensor networks. *Medium Access Control in Wireless Networks*, vol. 2, pages 535–560, 2008.

44 A. Bachir, M. Dohler, T. Watteyne, and K. K. Leung. MAC essentials for wireless sensor networks. *Communications Surveys & Tutorials, IEEE*, vol. 12, pages 222–248, 2010.

45 K. Langendoen. *MAC Alphabet Soup.* Available: http://www.st.ewi.tudelft.nl/~koen/MACsoup/.

46 H. Pei, X. Li, S. Soltani, M. W. Mutka, and X. Ning. The evolution of MAC protocols in wireless sensor networks: a survey. *Communications Surveys & Tutorials, IEEE*, 15:101–120, 2013.

47 F. Dobslaw, M. Gidlund, and Z. Tingting. Challenges for the use of data aggregation in industrial wireless sensor networks. In *Automation Science and Engineering (CASE), 2015 IEEE International Conference on*, 2015, pages 138–144.

48 M. Raza, S. Hussain, H. Le-Minh, N. Aslam. Novel MAC Layer Proposal for Ultra-Reliable and Low-Latency Communication in Industrial Wireless Sensor Networks; A Proposed Scenario for 5G Powered Networks. *ZTE Communications* 2017.

49 A. A. Kumar Somappa, K. Øvsthus, and L. M. Kristensen. An industrial perspective on wireless sensor networks – a survey of requirements, protocols, and challenges. *Communications Surveys & Tutorials, IEEE*, 16:1391–1412, 2014.

3

Haptic Networking Supporting Vertical Industries

Luis Sequeira, Konstantinos Antonakoglou, Maliheh Mahlouji and Toktam Mahmoodi

Touch is currently seen as the modality that will complement hearing and vision as a third media stream over the Internet in a variety of future haptic applications which will allow full immersion and that will, in many ways, impact society. Nevertheless, the high requirements of these applications demand networks which allow Ultra-Reliable and Low-Latency Communication (URLLC) for the challenging task of applying the required Quality of Service (QoS) for maintaining the user's Quality of Experience (QoE) at optimum levels.

In this chapter, we introduce the basics for haptic communication and discuss its QoS and QoE requirements as well as the main Key Performance Indicators (KPIs) for the Tactile Internet. In addition, we review and evaluate notable methodologies and technologies of bilateral teleoperation stability control and haptic data reduction, necessary components of the haptic communication infrastructure. Furthermore, we take into account the 5th Generation (5G) architecture, KPIs and some 3rd Generation Partnership Project (3GPP) components as well as the European Telecommunications Standards Institute (ETSI) Network Function Virtualisation (NFV) management and orchestration capabilities that will allow haptic applications to take life, in combination with the haptic data communication protocols, bilateral teleoperation control schemes and haptic data processing needed.

3.1 Tactile Internet Use Cases and Requirements

In teleoperation systems for haptic telemanipulation, otherwise known as *telehaptic* systems, usually, on one end of a communication channel a human operator is using a haptic interface (master device) to control a teleoperator device (slave actuator) on the other end. Teleoperation applications can be short-distance, where the communication channel can be wired or wireless without the need for network infrastructure, or long-distance, benefitting from packet-switched network infrastructures and transmitting their data as packets. The size of the communication channel for long-distance teleoperation can range from local area networks to the Internet.

Enabling 5G Communication Systems to Support Vertical Industries, First Edition.
Edited by Muhammad Ali Imran, Yusuf Abdulrahman Sambo and Qammer H. Abbasi.
© 2019 John Wiley & Sons Ltd. Published 2019 by John Wiley & Sons Ltd.

The goal of teleoperation systems is to provide to the user the feeling of presence in the remote environment where the teleoperator exists. It is a goal which can be achieved due to the ongoing improvement of the relevant hardware and software for providing the human users with multi-modal (visual, auditory and haptic) feedback. For this to be achieved, we need to meet the appropriate QoS requirements in order to maximize the QoE of the user.

3.1.1 Quality of Service

Internet-based Telepresence and Teleaction (TPTA) systems implement closed-loop control schemes over a real-time communication framework that allows interaction between a human operator and a remote environment using sensors and actuators. A system which can be described in this way is referred to as a Networked-Based Control System (NBCS) [1].

Since one of the core modules of such systems is the communication channel, several network protocols have been used or created for all teleoperation frameworks mentioned in previous chapters and their implementations over the Internet, supporting their efficient functionality either for virtual environment applications [2] or physical systems.

As with every networking system, the correct functionality of NBCS is liable to several obstacles which negatively affect their performance and can also be viewed as performance indicators which allow the comparison between different protocols and the quantification of the QoS they can deliver, which is of critical importance to applications such as telesurgery. Considering that teleoperation systems are NBCSs, it is natural to inherit these performance aspects with respect to requirements of TPTA systems. Common performance parameters are [3]:

- *Network delay* is the average time needed for a packet to travel from the input of the communication channel to its output. A survey that thoroughly lists and discusses the main sources of network delay as well as the solutions that can be currently implemented is [4].
- *Jitter* is the result of the influence delay has in packets independently, formally known as Packet Delay Variation (PDV), which affects the packet sequence. A common way to avoid jitter is to use packets with sequence numbers or timestamps; nonetheless this presumes the use of buffers, which in turn will increase the overall delay of the communication.
- *Packet loss* is a consequence of the network traffic congestion. As a result, the master and slave side of a TPTA system need to operate even with lack of information due to missing packets. Ways to overcome packet loss are substituting the missing values with null values, holding the last value or using interpolation (e.g. using a prediction method).
- *Data rate* of the communication channel. This can be affected by the sampling frequency, the sample resolution and the protocol overhead.

Additionally, alternative factors that may affect the system performance are signal quantization and other sources of noise.

The effects of packet loss with and without latency in discriminating visual and haptic events, although primarily focussing on packet loss, in the communication channel have been explored in [5], showing that both factors have an additive behaviour. Nonetheless, according to [6] the effects of packet loss can be managed by using mechanisms which increase the communication reliability, but as a result these mechanisms will increase the total latency. It is therefore a matter of balancing the trade-off between reliability and delay.

3.1.2 Use Cases and Requirements

It is obvious that the 5G network capabilities are determined by the requirements of the use cases which will need to utilize effectively the network. Essentially, we need to iterate through the use cases and extract those requirements. This is something that has already been done by 3GPP mainly in [7] with further information in [8]. In Table 3.1 we enlist the use cases using the first classification used in [7] along with a number of examples and briefly showing the main KPIs and requirements that need to be satisfied for the users to have good QoE.

Teleoperation is mainly related to the first three use case categories, but since a broad spectrum of applications exists, the different requirements can be grouped into many different classes. As seen in Table 3.1, the names of the use case families are self-descriptive as they include some of the main KPIs mentioned or a combination of them. These KPIs were selected to better demonstrate the main similarities and differences among the use cases. These include:

- *End-to-end latency* (e2e latency): The time it takes for data to be transferred from source device to destination (in milliseconds).
- *Reliability*: The number of packets successfully received by one end node divided by the total number of packets sent (percentage).
- *Availability*: The amount of time the communication system can provide service to the user divided by the total amount of time which is expect to deliver the services.
- *Mobility*: The speed at which the user is requesting services from the network provider. One example is telesurgery with the patient inside an ambulance moving at high speed.
- *Data rate*: The amount of data that the network can deliver in one second.
- *Coverage*: The area in which a network provider can offer services.
- *Positioning accuracy*: The accuracy at which a user's location can be tracked.
- *Security*: Maintaining the integrity of the data, in many cases, is a basic requirement. In Table 3.1, we also mention the relevant concept of confidentiality, which also relies on the network operator's discretion.
- *Service continuity*: Even when there is a change in the way a service is delivered to the user, this needs to happen in a seamless manner. This change can be a different access technology (e.g. satellite).
- *Energy efficiency*: The amount of bits per joule of energy consumed.

We need to mention that most of these KPIs behave differently depending on whether the user is in an indoor or an outdoor environment.

Evidently, in 5G networks, QoS provision will have to dynamically change according to the requirements of each of the Tactile Internet use cases. Apart from improvements in 5G technologies and physical network components, a flexible and scalable architecture is also required. Furthermore, in the 5G Core Network and Radio Access Network

Table 3.1 Classification 5G use cases with examples and their corresponding requirements for the main KPI.

Use case family	Traffic scenario examples	Main KPIs	Requirements
Higher reliability, availability & lower latency	• Medical treatment in ambulance • Low-latency industrial applications • Telemedicine cloud applications	e2e latency Reliability Availability Mobility Data rate	≤ 1 ms $\geq 99,999\%$ $\approx 100\%$ ≥ 120 km/h 10s of Mbps per device
Very low latency	• Human interaction, Immersive VR, Remote healthcare, Telementoring	e2e latency	1 ms one-way
Mission-critical services	• Prioritized access when: the network is congested, simpler access procedures or guaranteed QoS are needed	e2e latency Reliability Security	down to 1 ms $\approx 100\%$ max. confidentiality & integrity
Higher reliability & lower latency	• Unmanned Aerial Vehicles (UAVs) & Ground-based Vehicles • VR/AR applications • Cloud robotics • Industrial applications/Power plants	e2e latency Reliability Data rate Energy efficiency	1 ms min. 99,999% 250 Mb/s max. Various or NA
Higher accuracy positioning	• Outdoor positioning (high speed moving) • Indoor/Outdoor positioning (low speed moving) • UAV positioning for critical applications	Accuracy e2e latency Mobility	≤ 3 m for 80% of occasions $\leq$10 ms to 15 ms two-way ≈ 280 km/h
Higher availability	• Secondary connectivity for emergencies (mobile-to-satellite)	Coverage	Service continuity

(RAN), hardware and software are becoming clearly separated pieces of the architecture, which means that many 5G functionalities (Cloud RAN and Core Network) are virtualised and delivered as virtual machines or containers, creating network splits. That software separation allows certain network functions to be replaced with advanced features in less time or certain functionalities to be improved more easily and without having to change firmware or hardware devices. Software components can be migrated to place them as per need and they can also be scaled to provide more flexibility and scalability. In the NFV framework, cloud infrastructures provide a platform to run network functions, which usually consist of a combination of physical and software elements. Cloud technologies have became tools to allow a dynamic deployment of NFVs. In this context, they have to be able to interact with different technologies (e.g., 4th Generation

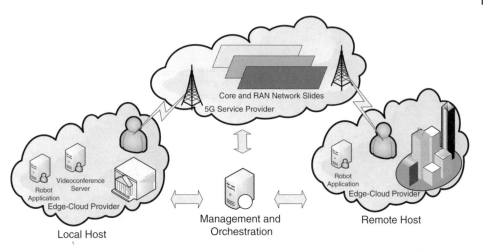

Figure 3.1 Use case scenario.

(4G), 5G and Internet of Things (IoT) systems) and to provide mechanisms for management and orchestration in a multi-systems environment ensuring interoperability across all.

For instance, Figure 3.1 depicts a cloud robotics scenario. The human operator and teleoperator will have a private cloud or an edge-cloud provided by a third party (both local and remote) and an external connection to the Internet, which can potentially be implemented by a 5G access network. In this context, a multi cloud environment needs to be managed to guarantee cloud inter-connectivity as well as network service orchestration across the different clouds. Furthermore, several functionalities, which the teleoperation equipment is responsible for, can be offloaded by the edge-cloud providers by bringing intelligence and processing power closer to the end nodes of the haptic communication system so as to provide adequate QoS.

3.2 Teleoperation Systems

3.2.1 Classification of Teleoperation Systems

Nowadays, extensive research has been carried out in bilateral and multilateral telehaptic systems. We can classify such systems based on the different communication delays and interaction levels that a user may experience. This results in two main categories, *Direct control* systems and *Supervisory control* systems, as described in [9]. Furthermore, as shown in Figure 3.2, there are three subcategories:

1. *Direct control*: The human operator interacts in real time with the environment while the master and slave devices communicate using position/force signals.
   (a) *Closed-loop with negligible delay*: Minimum delay in the communication channel and therefore the user is restricted to being in close proximity to the slave device.
   (b) *Time-delayed closed loop*: The most common form of teleoperation for digital closed-loop control systems. Similarly to the previous subcategory, the master device controls the slave actuator but the user is less restricted in terms of

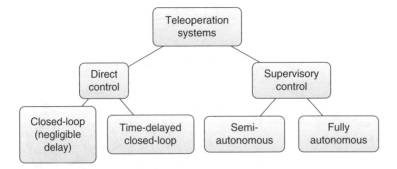

Figure 3.2 A classification of teleoperation systems.

distance from the device (e.g. transatlantic teleoperation). The remote side is not autonomous; however, an internal control loop which processes the command signals from the master device is included in the teleoperator. In this case, the communication channel (e.g. the Internet) may introduce variable delays [10].

2. *Supervisory control*: The teleoperator is
   (a) autonomously or
   (b) semi-autonomously
   controlled and receives high-level commands from the master. This is also referred to as "task-based teleoperation". Examples are teleoperation across planets or teleoperated robots with autonomous functionalities [11].

The three main solution spaces to improve haptic communication are (see Figure 3.3):

- The *communication network* solution space, covering both aspects of the Internet and the mobile/wireless communication that enables the Tactile Internet.
- *Data processing* solutions to reduce data transmission using perceptual thresholds or prediction methods.
- *Stability control* solutions to reduce the effect of extra delay and provide stability for the control loop of the haptic system.

Improvements in all solution spaces of haptic communication are under development and research. Main contributions to these solution spaces will be presented in the next chapters. Individual or joint improvement of the communication channel, control components and signal processing will guarantee high teleoperation quality, system stability and scalability. While current research studies address mainly these solution spaces independently (few studies address two of these spaces jointly), the ultimate solution for enabling the haptic communication should be based on joint optimization of these three solution spaces.

3.2.2 Haptic Control and Data Reduction

The ability to manipulate remote environments is one of the most interesting and also challenging tasks enabled by URLLC, which is an important feature of the new mobile generation i.e. 5G. More specifically, teleoperation is considered as a Critical Machine Type Communication (C-MTC) application of 5G, which requires ultra-low latency while maintaining high reliability.

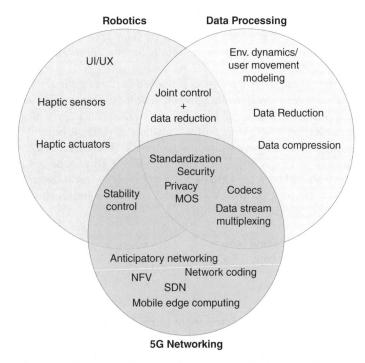

Figure 3.3 The main challenges in haptic communication over 5G network.

Figure 3.4 A teleoperation system block diagram [12].

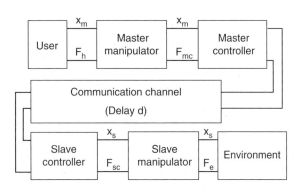

A schematic of teleoperation system is depicted in Figure 3.4. A "user" (in the master side) applies a force F_h to the "master manipulator" and there is also a "master controller" to control the output of the master manipulator (it might be master manipulator's position, velocity, etc.). Haptic data such as force and velocity is transmitted from master to slave and vice versa via communication channel. This communication channel introduces delay, jitter and packet drop to the haptic data. Therefore, the receive side gets the delayed and erroneous version of the actual haptic data. Similar to the master side, in the slave side, there is a "slave manipulator" that has some interactions with the "environment" and also there is a "slave controller" that controls the output of the slave manipulator.

If the teleoperation system is being controlled in a closed loop, even a small amount of delay in the loop causes instability and therefore it is necessary to use an ultra-low latency communication channel for these kinds of applications. However, different control schemes for teleoperation have different sensitivity to delay and knowing this sensitivity might help us in choosing the best configuration for the communication channel.

In this section, we will review some control schemes proposed in the literature. Each control scheme has some advantages and disadvantages and in order to choose a proper method for a particular application, one should pay close attention to information available, amount of communication delay, delay profile (constant or variable, bounded or unbounded), desired performance, environment structure and implementation limitations. Therefore, there is no best scheme that can be used for all teleoperation applications and maybe there should be some kind of negotiation between control and communication parts to find the best control scheme and communication channel configuration for a specific application and/or condition.

3.2.2.1 Performance of Teleoperation Control Schemes

In a teleoperation system (depicted in Figure 3.4), the relationships between forces are as follows [12]:

$$F_m = F_h - F_{mc}, \quad F_s = F_e + F_{sc}, \tag{3.1}$$

where F_h is human (user) force and F_e is the environment force. F_{mc} and F_{sc} are master and slave controller forces which are determined based on the control scheme.

Here, we will compare some control schemes and provide a summary of each method and then we will discuss their pros and cons. We have aimed to include as many control schemes as possible, as they cover most types of other methods. The control schemes that we considered are Force Reflection (FR), Shared Compliance Control (SCC), Force Reflection with Passivity (FRP), Position Error (PE), Four Channel (4C), Wave Variable (WV), Adaptive Motion/Force Control (AMFC), Passivity Controller (PC), Predictive Controller with Passivity (PCP) and Sliding Mode Controller (SMC).

There are some criteria that help us to analyse performance of different control schemes. Based on the application, some of these criteria might be more important than the others. In this section, we are more interested to know whether communication channel delay, denoted by d, should be known as an input to the control scheme or not. Another interesting characteristic is sensitivity of *stability*, *transparency* and *tracking precision* criteria of each scheme to the amount of delay – in other words, which criteria would be more affected by changing communication delay (d). The criteria suggested by [12] and [13] are as follows.

1. *Stability* of the closed loop teleoperation system
2. *Inertia (M_p) and damping (B_p)* perceived by the human when no force is applied by the environment to the slave manipulator
3. *Stiffness (K_p)* perceived by the human when a force is exerted to the slave manipulator by the environment
4. *Tracking error (δ)* of master manipulator position by the slave manipulator when no force is applied by the environment to the slave manipulator
5. *Position Drift (Δ)* when a force is exerted to the slave manipulator by the environment

The first criterion is stability, which is a necessary requirement for a control scheme to provide and is highly dependent on delay value [14]. A stable system can be Intrinsically

Stable (IS), which means that it is stable for all communication delay (d) and all control parameters, or it can be Possibly Stable (PS), which means that it is stable for all delays for specific parameters or is stable for delays less than d_{max} for some parameters [12].

M_p, B_p and K_p represent *transparency* or, in other words, quality of telepresence of human from the master side in the remote slave side. To have ideal telepresence, perceived inertia and damping in the case of no environmental force should be near the slave side inertia, damping and stiffness. To have a flawless manipulation – in other words, *tracking precision* – tracking error and position drift should be as low as possible. For more about stability and transparency of bilateral teleoperation system see [10].

In [12, 15] and [13] identical mass-damper dynamic is considered for both master and slave manipulators and a spring-damper model is considered for the environment as:

$$F_m = (M_m s + B_m)v_m \tag{3.2}$$
$$F_s = (M_s s + B_s)v_s, \tag{3.3}$$

where M_m and B_m are mass and viscous coefficients of master manipulator and M_s and B_s denote mass and viscous coefficient of the slave manipulator. In addition, it is assumed that master and slave manipulators are identical i.e. $M_m = M_s$ and $B_m = B_s$. Then, for each of the following control schemes, a formula is provided for "Master Controller" and "Slave Controller" blocks in Figure 3.4. Then equations for the above criteria are extracted. We use those equations to analyse and compare control schemes based on our criteria, which are, as we said before, requirement to know delay and sensitivity to the amount of delay. However, we aim to discuss all features and limitations of each method as much as possible.

Performance of Control Schemes   In order for a teleoperation system to work properly, not only stability must be guaranteed, but also transparency and tracking precision must be acceptable. In other words, stability is a vital property that must be ensured in every teleoperation system, but ensuring stability is not enough to accomplish the teleoperation task and in addition to stability, transparency and tracking precision need to be provided. Based on transparency and tracking precision parameters (i.e. inertia, damping and stiffness perceived from the master side as well as tracking error and position drift), a control performance loss function could be deployed in order to guarantee a good telepresence quality in the remote side.

Different control schemes have different orders of dependency to the communication delay, as shown in Table 3.2. Then, different control schemes can be compared to find the best one depending on the communication channel and teleoperation conditions. With this information, the communication channel parameters can be tuned to guarantee good performance for a specific application. In the paragraphs below, we will discuss the control schemes listed in Table 3.2 as well as another mode of teleoperation called the Model-Mediated Teleoperation Approach (MMTA).

Force Reflection (FR)   In the FR or Kinesthetic Force Feedback (KFF) scheme which uses Position-Force (PF) architecture, position is transmitted from master to slave in order to control the slave position and also force is transmitted from slave to master in order for the master to have a sense of the slave environment and to adapt the position of master manipulator based on the force received from the remote environment ([16–18]).

Table 3.2 Performance of control schemes based on delay.

Scheme	Stability	Known d	M_p	B_p	K_p	δ	Δ
FR	PS	No	$\propto d$	-	-	$\propto d$	-
SCC	PS	No	$\propto d$	-	-	$\propto d$	-
FRP	IS	No	$\propto d$	-	-	-	-
PE	PS	No	$\propto d, d^2$	$\propto d$	-	-	-
4C	PS	No	$\propto d$	-	$\propto 1/d$	-	$\propto d$
WV	IS	No	$\propto d$	-	$\propto 1/d$	$\propto d$	$\propto d$
AMFC	PS	No	$\propto d$	$\propto d$	-	-	-
PrC	PS	Yes	-	-	-	$\propto d$	-
PCP	IS	Yes	-	-	$\propto 1/d$	$\propto d$	$\propto d$
SMC	PS	No	-	-	-	-	-

This scheme is maybe the oldest and the simplest control scheme for teleoperation [13]. In this scheme, the "Master Controller" block in Figure 3.4 is the delayed version of F_{sc} (because of transmission through communication channel) multiplied by a control parameter (gain). The "Slave Controller" block in Figure 3.4 is another control parameter (gain) multiplied by the error i.e. the difference between delayed position of master manipulator (x_{md}) and the slave manipulator position (x_s). In this way, the human operator can feel the force exerted on the slave manipulator by the environment and intuitively adapt the applied human force (F_h) to achieve the desired position. However, in the presence of communication channel delay, the force feedback might destabilize the system even in smaller delays compared to the open loop scenario when no force feedback is available in free space ($F_e = 0$), since the delayed feedback might mislead the human operator and also the slave controller. Nevertheless, having force feedback is beneficial since we are interested to have telepresence as well as a stable system. The following (SCC and FRP) schemes may improve the stability of the FR scheme.

Based on [13], FR is PS and if we choose the slave controller gain sufficiently low, we will make the teleoperation system stable for large delays and at the same time we will lose transparency and tracking precision. In fact, the slave controller gain in this scheme is chosen in a compromise between stability and performance (transparency and tracking precision). In this method, we have to know neither the amount of communication channel delay nor the master and slave model. In terms of sensitivity to delay, perceived inertia and tracking error are linearly proportional to the amount of communication delay (d). However, position drift, perceived damping and perceived stiffness are independent of d.

Shared Compliance Control (SCC) The SCC scheme is similar to the FR scheme and it also uses PF architecture. The only difference is that a compliance term is added to the previous slave controller to comply with the position change caused by the environment force (F_e). In fact, F_e is passing through a low pass filter ($G(s) = \frac{K}{1+\tau s}$) which estimates the position change caused by F_e by modelling the environment as a spring damper and it aims to soften sudden changes in response to the sudden environment force. This term is then added to the difference between delayed master position and slave position i.e. $x_{md} - x_s$. The total output is multiplied by the slave controller gain, which can be tuned to get the desired performance. For more information about SCC see [18, 19]. It is interesting to note that "shared" refers to the fact that both human and environment control the slave manipulator [18].

In [18] an experimental comparison between FR and SCC for a 6 Degrees of Freedom (DoF) teleoperation system has been done based on completion time and integral of square forces and SCC showed better results in the presence of higher delays, since in SCC there is an internal loop in the slave side that would not be affected by the communication channel delay (d). Based on these observations, the paper suggests that in higher delays, only the internal loop (compliance loop in the slave side) should be used and the force reflection to the master should be removed, while in lower delays, both internal and external loops are helpful for transparency and stability.

Based on [13], SCC is PS but according to [18], compared to FR scheme, SCC improves stability in the presence of communication delays because of the internal loop in the slave side. In this method, neither do we have to know the amount of communication channel delay nor the master and slave model. However, knowing the amount of delay can help if we want to remove the force feedback to the master in higher communication delays.

In terms of sensitivity to delay, perceived inertia and tracking error are linearly proportional to the amount of communication delay (d). Position drift, perceived damping and perceived stiffness are independent of d. For SCC, perceived inertia, perceived damping and tracking error are exactly the same as FR, but FR has superiority to SCC in terms of perceived stiffness and position drift. Therefore, SCC improves stability while sacrificing transparency and tracking precision.

Force Reflection with Passivity (FRP) By adding a damping element into the FR scheme, passivity can be guaranteed, and by using a Lyapunov function stability is proved in the case of no external input to the system, and if the system is strictly passive, it will be asymptotically stable [20, 21]. The frequency domain controller equations in this scheme are as follows [12]:

$$F_{mc} = K_m F_{sd} + b v_m \tag{3.4}$$

$$F_{sc} = K_s \left(\frac{v_{md} - F_{sc}/b}{s} - x_s \right), \tag{3.5}$$

where v_{md} and F_{sd} represent delayed versions of v_m and F_s respectively. b is a damping impedance; however, this element will cause position drift in the slave while it keeps tracking master velocity.

Based on [13], FRP is IS, inertia perceived in the master side is linearly proportional to the communication channel delay (d) and stiffness perceived at the master side is zero, which is not good, and perceived damping is independent of d. However, tracking error and position drift tends to be really large. In this method, we do not have to know the amount of d and also we do not need to know the master and slave model to design the controller.

Position Error (PE) In the PE scheme, which uses Position-Position (PP) architecture [22], position is sent from master to slave and visa versa. Then, the position differences between master and slave manipulators multiplied by a control parameter (gain) will be used in "Master Controller" and "Slave Controller" blocks in Figure 3.4. However, due to communication delay, the difference between master position (x_m) and *delayed* slave position (x_{sd}) is available in the master side and the same for the slave side i.e. x_s and x_{md} are available. Therefore, large communication delay will destabilize the system and also will adversely affect transparency and tracking precision.

In [23], a transparency comparison between four possible architecture of two-channel teleoperation (Force-Position (FP), PF, PP, Force-Force (FF)) has been done and it shows that, neglecting communication channel delay, full transparency is not attainable in PP and FF architecture. Therefore, a PE scheme which is based on PP architecture is not as good as the previous FR methods (with PF architecture) in terms of transparency.

However, based on [13], stability improves in PE compared to FR as there is a trade-off between transparency and stability of teleoperation system. However, PE is still PS. In addition, in a PE scheme, inertia perceived in the master side is proportional to the square of the communication channel delay (d) and damping perceived is linearly proportional to d. Perceived stiffness, position drift and tracking error are independent of d. In this method, we do not have to know the amount of d and the master and slave model.

Four Channel (4C) The 4C control scheme ([10, 24]) is based on the assumption that force and velocity information are transmitted from master to slave side and also from slave to master side. Therefore, we will have four information channels in total. 4C equations in frequency domain for master and slave controllers are as follows [12]:

$$F_{mc} = -C_6 F_h + C_m v_m + F_{sd} + v_{sd} \tag{3.6}$$

$$F_{sc} = C_5 F_e - C_s v_s + F_{md} + v_{md} \tag{3.7}$$

$$v_{md} = C_1 e^{-sd} v_m, \quad F_{md} = C_3 e^{-sd} F_m \tag{3.8}$$

$$F_{sd} = C_2 e^{-sd} F_s, \quad v_{sd} = C_4 e^{-sd} v_s. \tag{3.9}$$

In [25], control parameters C_1, C_2, C_3 ans C_4 are designed to achieve full transparency as:

$$C_1 = Z_s + C_s, \quad C_2 = 1 + C_6 \tag{3.10}$$

$$C_4 = -(Z_m + C_m), \quad C_3 = 1 + C_5 \tag{3.11}$$

$$Z_s = M_s s + B_s, \quad Z_m = M_m s + B_m \tag{3.12}$$

$$(C_2, C_3) \neq (0, 0). \tag{3.13}$$

In fact, C_5 and C_6 represent local force feedback loops in master and slave side proposed in [25] to improve stability and performance (transparency and tracking precision). As Lawrence concluded in [10], in ideal conditions (no communication channel delay), neither PP nor PF architecture (i.e PE and FR control schemes) are able to provide high transparency and stability; however, FRP (passivated PF architecture) improves passivity (stability) but provides low transparency improvement. Obviously, and as shown in [10], the use of all 4C information can help to provide better transparency and stability if the control parameters are optimized in a trade-off between the two.

Based on [13], 4C is PS, inertia perceived in the master side and position drift are linearly proportional to the communication channel delay (d) and damping perceived is zero. Perceived stiffness is inversely proportional to d and tracking error is independent of d. In this method, we do not have to know the amount of d but we have to measure the master and slave model to design the controller in Equation (3.15).

Wave Variable (WV) One idea to guarantee stability of a teleoperation system is to maintain its passivity since a passive system is proved to be stable. Intuitively, according to the definition of a passive system, which is a system in which output energy is equal to or less than input energy, one can immediately conclude that a bounded input to this system results in a bounded output.

The WV control scheme is designed based on the idea that stability of the teleoperation system can be maintained in varying time delay by using wave variable filters [26]. Therefore, by transmission of wave variables instead of force and velocity (the actual physical variables), the communication delay would become passive because of storing the input wave energy for the duration of delay and in this way the delay block would become lossless and passive regardless of its value or phase lag, while normally delay in the system is an active element. The proof of this also is presented in [27].

The schematic of wave variable control scheme is shown in Figure 3.5 and the wave variables S_m^+, S_m^-, S_s^+, and S_s^- are as follows:

$$S_m^+ = \frac{bv_m + F_m}{\sqrt{2b}}, \quad S_m^- = \frac{bv_m - F_m}{\sqrt{2b}} \tag{3.14}$$

$$S_s^+ = S_m^+ e^{-sd}, \quad S_s^- = S_m^- e^{+sd} \tag{3.15}$$

The transparency and tracking precision parameters for our mass damper control system with the Proportional-Derivative (PD) controller in the slave side (with parameters B and K in Figure 3.5) of the wave variable scheme are as follows:

$$M_p = 2M_m + bd - \frac{B_m^2}{K} - \frac{B_m^2}{b}d \tag{3.16}$$

$$B_p = 2B_m \tag{3.17}$$

$$K_p = \left(\frac{1}{K} + \frac{d}{b} + \frac{1}{K_e}\right)^{-1} \tag{3.18}$$

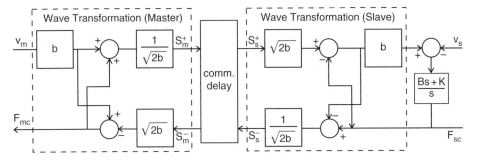

Figure 3.5 Wave variable control scheme for teleoperation system.

$$\delta = \frac{d}{2b} + \frac{1}{2K} \tag{3.19}$$

$$\Delta = \frac{1}{K} + \frac{d}{b}. \tag{3.20}$$

As can be seen in the above formulas, the cost that we have to pay to guarantee stability in this scheme is losing transparency and tracking abilities, since all the above parameters except B_m are dependant on the communication delay (d), so compared to other schemes it has more dependency on d (the comparison is summarized in the Table 3.2). It means that although we have compensated for the effect of delay on the stability, we have passed it on to the transparency and tracking parameters.

THe WV control scheme, guarantees stability of the teleoperation system through passivity and therefore it is IS. In this scheme, there is no need to know the communication delay or the environment model and we only need to transmit and receive wave variables instead of the actual physical variables. In the WV scheme, inertia perceived in the master side, tracking error and position drift are linearly proportional to the communication channel delay (d), perceived stiffness is inversely proportional to d and perceived damping is independent of d [13].

Adaptive Motion/Force Control (AMFC)   Based on the review of adaptive control schemes for teleoperation systems that is presented in [28], two main approaches are considered: 1) adaptive-gain control 2) disturbance accommodating adaptive control. Basically, in adaptive control, some coefficients are automatically being adjusted online to get the desired performance.

According to the adaptive-gain control mechanism proposed in [29] and [30], performance (transparency and tracking precision) criteria are computed in [13] and, based on that, AMFC is PS, inertia perceived in the master side is linearly proportional to the communication channel delay (d). Perceived damping and stiffness, tracking error and position drift are independent of d. In this method, we do not have to know the amount of d but we have to estimate the master and slave model to design the controller.

Predictive Controller (PrC)   In a Predictive Control (PrC) scheme [31], the slave and remote environment model is used to *predict* the current status of the slave side based on its available delayed version in order to remove the delay term from the denominator of the closed loop transfer function. PrC provides stability and better performance in larger delays compared to other methods that are based on passivity and robust control techniques [32].

Smith Predictor [33], is one of the oldest predictive control mechanism which can be considered for delay compensation in teleoperation systems. In [32], four possible architectures of two-channel architecture are considered and among them PF architecture is chosen because of better transparency. A block diagram of this scheme is shown in Figure 3.6 in which a Smith Predictor is implemented on the master side and a PD controller is used for the slave side. The corresponding equations are:

$$F_{mc} = F_{sd} + (1 - e^{-2sd})\frac{(Bs + K)(M_s s^2 + B_s s)}{M_s s^2 + (B_s + B)s + K}x_m \tag{3.21}$$

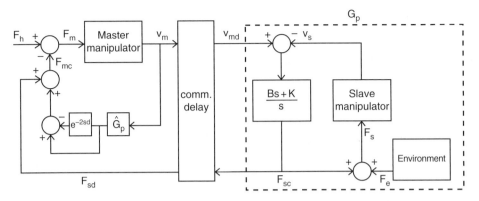

Figure 3.6 Smith Predictor with FR architecture [32].

$$F_{sc} = (Bs + K)(x_{md} - x_s) \tag{3.22}$$

$$F_{sd} = e^{-sd}F_s, \quad x_{md} = e^{-sd}x_m, \tag{3.23}$$

where B and K are PD control parameters.

In this control scheme, we have to know the slave model and also the amount of communication delay (d) and thus errors in system model and delay estimation cause problems and instability, which are discussed and analysed in [31, 34].

Based on [13], the PrC scheme is really similar to FR except that it predicts current force feedback with current master position to compensate for delay effect. PrC is PS and interaction with the environment and unknown F_e causes instability. In terms of transparency and tracking precision, inertia, damping and stiffness perceived as well as position drift are independent of d while tracking error is linearly proportional to d.

Predictive Controller with Passivity (PCP)   To improve the stability of the PrC scheme, [35] proposed a new control scheme which is a combination of prediction with wave variable and designed a regulator to guarantee its passivity. The advantage of this method compared to the PrC scheme is that it can tolerate higher delays without becoming unstable.

In this scheme, wave variables are transmitted on the communication channel between master and slave instead of power variables i.e force and velocity. However, it is impossible to transmit the environment power and this is one of the limitations of this method.

Based on [13], since this scheme maintains passivity, it will be IS. Performance (transparency and tracking precision) metrics of PCP are very similar to WV if the control parameters are chosen to have the best transparency and tracking precision. However, tracking error in PCP, like PrC, is linearly proportional to d. In summary, inertia perceived, tracking error and position drift are linearly proportional to d, stiffness perceived is proportional to $1/d$ and damping perceived is independent of d. In addition, we have to know communication delay and remote system model.

Sliding Mode Controller (SMC)   SMC has been widely used for bilateral teleoperation and is known to be robust to model uncertainties and variations. This property is useful to compensate for varying time delay ([36–39]). Moreover, a modified SMC is proposed in

[37] to overcome time delay variations due to using the Internet as the communication medium, since the nonlinear gain in this method is set independent of d.

Here, we consider the control scheme used in [37] which uses impedance control for master side and a SMC for slave side with these equations in the time domain ([37]):

$$F_{mc} = \left(\frac{M_m}{M} - B_m\right) v_m + \left(1 - \frac{M_m}{M}\right) - \frac{M_m}{M}(F_{ed} + Kx_m) \tag{3.24}$$

$$F_{sc} = B_s v_s - F_{e\cdot} - \frac{M_s}{M}(Bv_{md} + Kx_{md} - F_{hd} - F_{edd}) - M_s\lambda\dot{e} - K_{gain}\, sat\left(\frac{s_e}{\phi}\right)$$

$$e = x_s - x_{md}, \quad s_e = \dot{e} + \lambda e \tag{3.25}$$

$$v_{md} = v_m(t - d_1), \quad x_{md} = x_m(t - d), \quad F_{hd} = F_h(t - d_1) \tag{3.26}$$

$$F_{ed} = F_e(t - d_1), \quad F_{edd} = F_e(t - d_1 - d_2), \tag{3.27}$$

where M, B, and K are parameters considered for the impedance control, $sat(.)$ is the saturation function, λ is a strictly positive parameter, ϕ is the boundary layer thickness and s_d is the sliding surface (for more information see [37]).

One of the most important disadvantages of the regular SMC method is the chattering problem (oscillations with finite amplitude and frequency), which will get worse when K_{gain} is high. However, [37] claims that by using the modified SMC, K_{gain} boundary will become independent of the communication delay, and doesn't have to be set higher in larger delays.

Assuming d_1 and d_2 are identical and equal to d, based on analysis in [13], inertia, damping and stiffness perceived as well as position error are independent of d while tracking error is linearly proportional to d. The SMC control scheme is PS and its stability is similar to FR scheme. However, if $F_e = 0$, it will be IS. In this method, we have to know the master and slave model to design the controller but there is no need to know d. The environment force (F_e) should be transmitted from the slave to the master and also human force, master velocity (and maybe its position) and the transmitted environment force (from slave to master) should be transmitted again from the master to the slave.

Time-Domain Passivity Control The Time-domain Passivity Approach (TDPA) was defined by Hannaford and Ryu [40] for haptic interfaces and extended to apply to teleoperation systems [41]. The approach has gained interest during the past few years due to its simplicity and robustness to communication delays.

The basic concept behind time-domain passivity control is to monitor the energy flowing to and from the master, the slave side or both in real time using a Passivity Observer (PO) which can be placed in series or in parallel to the communication channel (see Figure 3.7). In the series arrangement we choose velocity as an input, whereas in the parallel arrangement force is used as input to the PO. If the PO decides that the passivity condition is not satisfied, meaning that the system generates energy and therefore is active, then a PC has the responsibility to retain the system's passivity by using adjustable damping elements [42].

Following a relevant arrangement in [43], after the acquisition of the environment parameters (related to velocity and force data of the slave device) and transmission through the communication channel, a model of the environment is created on the

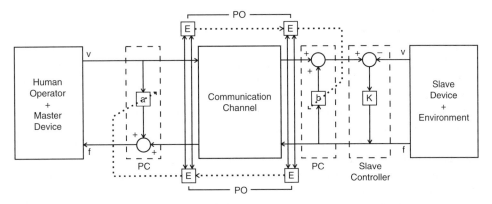

Figure 3.7 A bilateral teleoperation system using the time-domain passivity control architecture (from [48]). The PO entities compute the energy flows for both directions and provide input to the PC on each side.

master side and according to this model the damping coefficients of a PC are adjusted according to a PO's output.

Bounding energy signals [44] or control signals [45] of TDPA systems have showed improvement of the method's effectiveness. In [46], a different scheme is proposed, where the segment of the control system on the teleoperator side, including the communication block, is considered as a one-port network that receives position and provides force feedback. A method that combines time domain passivity control with perceptual data reduction is introduced in [47].

An augmented version of TDPA was proposed in [49] based on the framework of network-based analysis of passivity-based teleoperation systems in [50]. Modelling the teleoperation system using an electrical representation, rather than a mechanical one, is beneficial due to its simplicity. The electrical representation employs ideal flow (velocity) and effort (force) dependent sources as the analogous system elements to the motion commands of the human operator and reflected force of the teleoperator. These sources can also be delayed dependent.

The communication channel equivalent is called Time Delay Power Network (TDPN) and it is in the form of a two-port subsystem that can be coupled with a passivity controller. Another differentiation of this framework lies in the possible structures of the proposed architecture as it further disambiguates the network channel representation with regard to the energy flows.

Model-Mediated Teleoperation Approach To guarantee both the system stability and transparency at the same time in the presence of arbitrary communication delay, the concept of the MMTA has been proposed. Rather than directly sending back the haptic (force) signals, the parameters of the object model which approximate the remote environment are estimated and transmitted back to the master in real time during the slave's interaction with the remote environment. The model parameters include the surface geometry and physical properties of the remote objects. On the master side, a copy of this object model is maintained according to the received model parameters, and the haptic feedback is computed on the basis of the local model without any delay. The MMTA was first presented in [16] and afterwards extended in [51].

The MMTA opens the control loop between the master and slave and leads to two decoupled control loops, one on the master and one on the slave side. The stability of the MMTA system can be determined using the stability of the human-master local model closed loop and the slave-environment closed loop [52, 53]. If the estimated model is an accurate approximation of the remote environment, then both stable and transparent teleoperation can be achieved.

More specifically, in [16], the local control loops at one side of the haptic communication system aim to simulate the impedance observed at the opposite side. Later on, in [51], in contrast to transmitting position or force values, an abstraction layer was introduced, but implemented for a 1-DoF application. The suggested algorithm replicates the remote environment at the master side and issues commands through the communication channel to the slave device.

When the master receives new model parameters from the slave side, an update of the local model according to the received model parameters is required. Ideally, the parameters of the local model need to be updated to the correct ones as quickly as possible. However, improper update schemes, e.g., a sudden change in stiffness or model position, result in a suddenly changed force that is displayed to the human user. This is called the model-jump effect [54]. To allow for a moderate model update which guarantees the stability, passivity-based schemes were developed [51, 54, 55]. In [51, 54], the model position is updated only if no energy is injected into the local model system after the update. In [55], the authors used an adaptive damper to dissipate the energy injected into the local model system during the model update. This allows for a quicker model update and a higher subjective preference rate compared to the scheme proposed in [51, 54].

In general, the MMTA has the benefit of being simultaneously stable and transparent in 1D or simple 3D real environments compared to the passivity-based control approaches. However, due to the limitations of existing online model-estimation algorithms, the MMTA cannot work efficiently in complex or completely unknown environments.

There is no doubt that obtaining a precise object model for complex environments (both object geometry and physical properties) is the most important task and also the main challenge for the MMTA, since a perfect match between the local model and the environment enables stable and transparent teleoperation in the presence of arbitrary communication delays. Early attempts employ predefined model for MMTA systems [56]. This requires the master system to have rich knowledge about the remote environment. In practice, there are situations in which we have limited knowledge about the remote environment, especially when the slave enters a new environment or interacts with dynamic (movable or deformable) objects. Therefore, online environment modelling and model updating are inevitable.

Instead of modelling the environment, an alternative architecture of the MMTA is to model the behaviour of the human operator. The estimated model parameters on the master side are transmitted to the slave to guide the slave's motion. The slave is thus not controlled by the delayed master motion commands, but performs specific tasks in complete autonomy based on the received human behaviour model. Similarly, if the model as well as the model parameters can accurately approximate the human behaviour, the slave can behave like a human user and a complete skill transfer can be realized [57–59]. The modelling of human behaviour, however, is quite challenging and the model of human behaviour has not been fully studied yet. Most MMTA systems are thus based on the modelling of remote environments, but not the modelling of human behaviour.

The estimated model parameters need to be transmitted back to the master for building/updating the local model. The transmission happens normally when the slave enters a new environment, the environment the slave is interacting with changes, or the parameter estimation is not precise. Once the estimates converge to the true values, there will be no updates required and thus the system achieves zero transmission in the backward communication channel. For real teleoperation systems, however, the estimates can vary over time due to measurement noise, natural tremble of human arm movement, etc. Obviously, to transmit every estimate is a waste of the network resources.

Thus, an efficient data reduction scheme is needed to selectively transmit the estimated model parameters. Verscheure et al. [60] presented an event-triggered estimation scheme. The estimation and transmission are activated only when special conditions are satisfied, e.g. sufficiently large force/velocity of the slave, or sufficiently large displacement from the last estimation.

3.2.2.2 Haptic Data Reduction

Haptic sensor readings, especially when reading the kinesthetic signals, have a sampling rate of 1 kHz or even higher [10, 61, 62]. In order to keep the communication delay as small as possible, haptic samples are packetized and transmitted instantly. As a result, the communication of haptic data in teleoperation systems requires 1000 or more packets per second to be transmitted. For vibrotactile signals (touch emulated with vibrations), the sampling frequency greatly depends on the type of interaction of the user with the remote environment. For tasks of low precision, it requires a feedback frequency of 20 Hz to 30 Hz, while high-precision tasks require a feedback frequency of 5 kHz to 10 kHz [63–65]. Such high rate of packet transmission incur substantial data overhead due to the transmission of packet header and, thus, results in increasing latency [66, 67].

Future teleoperation systems which will be able to provide full body immersion will use a large number of sensors and actuators increasing proportionally to the number of DoF required by the haptic applications. Even though 5G networks will have data rate capabilities which can easily cover the needs of a haptic data transmission of a user, it is necessary to have in mind the additional transmission of audio and video data. Therefore, it is required to explore and improve haptic data reduction and compression methods.

Haptic data reduction techniques, either for kinesthetic data reduction or tactile data reduction, can be considered as lossy data reduction/compression schemes as full recovery of the original raw data is not possible. These techniques can be applied in the application layer since they rely on processing the data as acquired by the haptic devices. On the other hand, network throughput reduction can be achieved by other means such as Physical Layer Network Coding (PLNC) viewed from the scope of the 5G infrastructure. In this section, data reduction will be related to the processing of haptic data only.

3.2.2.3 Kinesthetic Data Reduction

Kinesthetic data reduction techniques are mainly based on two approaches of statistical and perceptual schemes [68]. The former one normally uses the statistics of the haptic signals to compress the packet size, while the latter one mainly focuses on reducing the packet rate over the communication network. Since the packet header overhead is significant as kinesthetic data packet payload size is small, reducing the frame rate seems to be an obvious choice for reducing the total amount of data.

Statistical Schemes Early attempts with respect to signal sampling employ predictive models to reduce data redundancy. Quantization techniques (e.g., Adaptive Differential Pulse Code Modulation (DPCM)) for kinesthetic data reduction are presented in [69]. In [70], kinesthetic data are 32-bit floating-point values. After the master and slave device have exchanged enough raw data, a simple position prediction method was proposed. Compression was achieved by performing an exclusive-or operation between the predicted and the previously predicted value and the result being reduced to 8 important bits.

Apart from prediction, lossy kinesthetic data compression and decompression has also been achieved by using Discrete Cosine Transform (DCT) [71], similarly to the Joint Photographic Experts Group (JPEG) codec, in a teleoperation system with force feedback with a compression ratio of 20%. Finally, another compression method that has been tested on 1-DoF haptic data is Wavelet Packet Transform (WPT) [72]. In this case, decompression is accomplished with the Inverse WPT (or IWPT).

Perceptual Schemes The first proposal that targets packet rate reduction for networked control systems can be found in [66]. In this work only samples that contain changes more than a given/fixed threshold are transmitted. The receiver reacts to a missing sample by holding the value of the most recently received sample. The approach in [66], however, ignores that the human operator comes with strong limitations in terms of perceivable signal changes.

State-of-the-art methods of perceptual data reduction have shown that it is possible to exploit the limitations of human operators and how they perceive haptic signals [73] towards achieving more efficient data reduction [74]. Such works mainly rely on a concept from psychophysics that is the difference threshold, otherwise known as Just Noticeable Difference (JND), which is the minimum amount of change in stimulus intensity needed for a perceptible increment in sensory experience. This threshold is formulated by Weber's law [75]:

$$\frac{\Delta I}{I} = c \tag{3.28}$$

where I is the stimulation intensity, ΔI is the difference of stimulation intensity to be perceived (the JND) and c is a constant, also known as the Weber's fraction. Difference thresholds, also known as discrimination thresholds, are defined both for haptic system parameters and quantities such as stiffness, velocity and force. These thresholds also differ depending on the movement scenario and the muscles involved [76]. For example, the JND when a human operator perceives force feedback to the index finger is approximately 10% [77].

Implementing kinesthetic data reduction showed up to 90% decrease in packet rate in [78]. Perceptual kinesthetic data reduction schemes have also been implemented for position and velocity signals using distance metrics (the Euclidean distance) between haptic data vectors (position vectors) [79]. This approach, however, needs further investigation because in psychophysics there is no result that shows Weber's law also applies to positions. This methodology also applies to orientation data and has also been extended to six DoF.

In comparison to other sampling methods such as the level crossings method (which incorporates absolute differences instead of percentages between samples), the perceptual-based kinesthetic data reduction schemes are proven to have good but similar accuracy [80]. Nonetheless, in [81] it is stated that the *level crossings sampler* outperforms the sampling method based on Weber's law. It has also been shown that the JND decreases with increase of the rate of kinesthetic force stimuli.

Perceptual Schemes with Predictive Coding   Prediction models of haptic signals can be used to estimate future haptic samples from previous data. This is able to achieve further reduction of haptic packet rate. The same predictors can run in parallel at both the master and slave sides. At the sender side, the predictor generates the predicted haptic signal at every sample instant. If the prediction error is smaller than the corresponding JND, no update is triggered. Otherwise, the input sample is transmitted to the other side and the transmitted sample is used for updating the prediction model. At the receiver side, if a packet is received, it is directly applied as the output and the received haptic signal is used for updating the prediction model. Otherwise, the predictor generates a predicted haptic signal as the current output.

The simplest but also the least efficient prediction method is the Zero-Order Hold (ZOH) predictor. When no data have been transmitted from the sender the receiver holds the last value of the sample it previously received.

Different kinds of predictors can be used to estimate the future haptic samples. For example in [67, 82], a linear predictor of the first order was adopted, namely First-Order Linear Predictor (FOLP). This simple predictor can lead to a significantly decreased packet rate up to 90–95% without deteriorating the immersiveness of the system. The velocity signal approximation used in the prediction model, however, is very sensitive to noise, even more than force signals. Therefore, haptic samples need to be filtered by using a low-pass filter to minimize the undesirable effects of measurement noise. An augmented version of this framework, presented in [83], introduces noise reduction by employing a scalar Kalman filter on the input signals.

Yet another prediction model that takes the prediction error into account was employed in [84] for three-dimensional position and force data, a third-order Auto Regressive (AR) model. According to the method's algorithm, after an initialization and training process, the adaptive coefficients of the model are computed so that the predicted values are produced. Afterwards, taking into account the JND threshold, the algorithm decides whether the training values need to be updated either from the predicted data or the current real data.

Contrary to transmitting sample values obtained from haptic devices or their derivatives, regression analysis also allows the transmission of only the model parameters. Such a method can be found in [85] where samples are first fitted in a quadratic curve.

Using a more complex predictor, a geometry-based prediction model was proposed in [86]. The remote environment is modeled either as a plane or a sphere according to the historical interaction, without taking friction, slave inertia and time delay of the network into account. The future haptic samples were predicted based on the interaction with the geometry model.

Taking into account a total of four predictors (ZOH, FOLP, plane and sphere predictors), a predictor selection method is also proposed in [86] which identifies the predictor with the least prediction error. Psychophysical tests on human subjects showed that the hybrid approach performed as well as the best predictor (sphere predictor).

Prediction of signal samples and perceptual coding of the signal was also proposed in [87] by implementing the Particle Filtering method. This framework uses the Probability Distribution Function (PDF) of the user's motion or force to predict future or lost samples.

3.2.2.4 Tactile Data Reduction

While data reduction on kinesthetic signals is widely investigated as discussed in the previous subsections, the number of studies on the compression of vibrotactile texture signals is limited. The kinesthetic signals involve large amplitude low-frequency force feedback and were found lacking in realism due to the absence of high-frequency transients (e.g., tapping on hard surfaces [88]) and small-scale surface details (e.g., palpation of textured surfaces [89]). Transmission of vibrotactile signals for increasing fidelity of real-time teleoperation systems and its storage for later playback necessitate data compression.

The necessary step before compressing the vibrotactile signals is to model them. Significant research has been devoted toward modelling tactile texture signals [90–92]. The vibrotactile signals are raised from coarse regularly patterned textures by decaying sinusoids [90]. In [91], Kuchenbecker et al. use a linear predictor to model the texture signals. In [92], the authors segment the recorded real-world vibrotactile texture signals based on their physical surface feature. These segmented signals are fitted and the corresponding filter parameters are stored. Then, a virtual visual-haptic model representing the previously extracted surface features is constructed and haptic rendering is performed based on this model. The above works, however, are not optimized for compression.

Towards the compression of vibrotactile signals, Okamoto and Yamada presented a frequency-domain texture compression algorithm loosely based on the knowledge of human vibrotactile perception [93]. The textured surfaces are scanned and the surface height is represented by a waveform. This waveform is transformed to the temporal frequency domain using the DCT, and the DCT coefficients are thresholded and quantized according to the knowledge of frequency-domain amplitude-JND for vibrotactile stimuli [94]. The authors of [93] showed a 75% compression of the texture data with guaranteed perceptual transparency. Unfortunately, this algorithm works only offline, which means prior knowledge about the surface must be known (e.g., pre-scanning procedure). Further research on JNDs of vibrotactile perception has been made in [95] by studying the JNDs with low-intensity reference stimuli, starting at 5 Hz, close to the sensory absolute threshold. In [96] three experiments were carried out, first an experiment showing that acceleration is not a vibration property that affects humans due to the nature of the human tissue, second an experiment to determine JNDs showing it is unimportant to subjectively optimize tactile displays and third an experiment that showed the impact of using different devices in low frequency vibrations (starting at 100 Hz).

The first online compression of vibrotactile signals can be found in [97] for bilateral teleoperation. The compression algorithm is inspired by the similarities observed between texture signals and speech signals. Thus, a well-developed speech coding technique, the Algebraic Code-Excited Linear Prediction Coding (ACE-LPC) [98], is adapted for developing a perceptually transparent texture codec. The authors of [97] reported a compression rate of 8 : 1 with a very low bitrate (4*kbps*) on data transmission. An extended version of this compress algorithm was proposed in [99], in which the masking phenomenon in the perception of wide-band vibrotactile signals was applied to further improve the efficiency of the texture codec. The masking phenomenon [100] implies that humans can tolerate larger errors in high-energy frequency bands, and smaller ones in low-energy frequency bands. Therefore, for encoding (compressing) the texture signals, the bitrate should be allocated more in the low-energy frequency bands compared than to the high-energy ones. In [99], the authors experimentally showed that the masking for haptics is very similar to its auditory analogue. With the help of the experimental results, the bitrate of the codec output can be driven down to as low as 2.3*kbps* without distorting the subjective perception.

Last but not least, it must be noted that there is currently no objective quality metric, such as Mean Opinion Score (MOS), for the evaluation of vibrotactile signals (with the exception of [101] on the effect of delayed kinesthetic and 3D video data on the user). Nonetheless, the similarities shared between audio and tactile signals will allow the design of tactile codecs in a similar fashion as with audio codecs [102].

3.2.3 Combining Control Schemes and Data Reduction

The aforementioned data reduction approaches for teleoperation systems have been initially developed without considering the stability issues and control scheme. In the presence of communication delays, however, the data compression schemes have to be combined with stability-ensuring control schemes. For this reason, joint control scheme and data reduction techniques have been proposed.

Haptic Data Reduction + Wave-Variable Control Architecture The Perceptual Deadband Method (PDM) packet rate reduction scheme has been combined with the WV control scheme in [103, 104] for dealing with constant communication delay. The PD approach is applied either on the wave variables [104] or on the time domain signals [103] (force and velocity). In order to modify the control schemes and to incorporate data reduction schemes, the passive PD schemes, such as the energy supervising transmission [104] and the passive ZOH reconstruction scheme [103], were developed. In [104], the authors experimentally found the subjectively best deadband parameter for interacting with a rigid wall. In contrast in [103], the authors showed that applying the PD approach on time-domain signals leads to better performance on both system transparency and data reduction compared to applying the PD approach on wave variables.

Haptic Data Reduction + TDPA Xu et al. [47] have recently combined the PD approach with the TDPA control scheme to reduce the packet rate over the communication network while preserving system stability in the presence of time-varying and unknown delays. On both master and slave sides the signals are processed with the deadband method to regulate the transmission rate of the velocity, force, and energy signals based on the PD approach discussed previously.

In order to incorporate the control scheme with the PD approach, the energy calculation in the PO is modified. At each sampling instant, if no update is received, the PO outputs the same energy as the most recently received one (ZOH reconstruction) for the subsequent computation.

Compared to the existing WV-based haptic data reduction approaches, the TDPA-based haptic data reduction scheme presented in [47] can robustly deal with time-varying delays and does not require the use of the passive PD approach. This is because the deadband controllers and reconstructors are set in between the two POs, and the TDPA is capable of ensuring passivity of any two-port networks between the POs on the master and slave side. Experiments show that the TDPA-based haptic data reduction scheme is subjectively more transparent compared to the WV-based schemes. In addition, it is able to reduce the packet rate by up to 80%, without significantly distorting user's experience for the tested communication delays of up to 100 ms ± 30 ms.

Haptic Data Reduction + MMTA   Similarly, a perception-based model update scheme is also incorporated into a MMTA architecture [105]. The environment model as well as its physical properties (stiffness and surface friction coefficient) are estimated at the slave side in real time and transmitted back to the master for building/updating the local model. The transmission happens normally when the slave enters a new environment, the environment the slave is interacting with changes, or the parameter estimation is not precise. Once the estimates converge to the true values, no updates are required and thus the system achieves zero transmission in the backward communication channel.

For real teleoperation systems, however, the estimates can vary over time due to measurement noise, natural tremble of human arm movement, etc. Obviously, to transmit every estimate is a waste of the network resources. Thus, an efficient data reduction scheme is needed to selectively transmit the estimated model parameters. Verscheure et al. [60] presented a event-triggered estimation scheme. The estimation and transmission are activated only when special conditions are satisfied, e.g. sufficiently large force/velocity of the slave, or sufficiently large displacement from the last estimation. Xu et al. [105] applied the PDM to the estimated model parameters to reduce the transmission rate. The authors also reported a data reduction of about 90% with guaranteed subjective quality of teleoperation.

The aforementioned perceptual or event-trigger control schemes for the MMTA avoid the transmission of irrelevant updates to reduce the packet rate on the network. System stability and transparency are verified in the presence of a round-trip communication delay of up to 1000 ms.

Acknowledgment

The authors are grateful for the financial support by the EPSRC for Initiate, and the EC for 5GPPP 5GCAR.

References

1 T.-C. Yang. Networked control system: a brief survey. *Control Theory and Applications, IEE Proceedings*, 153(4):403–412, July 2006. ISSN 1350-2379. doi: 10.1049/ip-cta:20050178.

2 Q.V. Dang, B. Allouche, A. Dequidt, L. Vermeiren, and V. Dubreucq. Real-time control of a force feedback haptic interface via ethercat fieldbus. In *Industrial*

Technology (ICIT), 2015 IEEE International Conference on, pages 441–446, 2015. doi: 10.1109/ICIT.2015.7125138.

3 S.G. Tzafestas. *Web-Based Control and Robotics Education*. Intelligent Systems, Control and Automation: Science and Engineering. Springer Netherlands, 2009. ISBN 9789048125050.

4 B. Briscoe, A. Brunstrom, A. Petlund, D. Hayes, D. Ros, I. J. Tsang, S. Gjessing, G. Fairhurst, C. Griwodz, and M. Welzl. Reducing internet latency: A survey of techniques and their merits. *IEEE Communications Surveys Tutorials*, 18(3):2149–2196, 2016. ISSN 1553-877X. doi: 10.1109/COMST.2014.2375213.

5 Z. Shi, H. Zou, M. Rank, L. Chen, S. Hirche, and H. J. Muller. Effects of packet loss and latency on the temporal discrimination of visual-haptic events. *IEEE Transactions on Haptics*, 3(1):28–36, Jan 2010. ISSN 1939-1412. doi: 10.1109/TOH.2009.45.

6 Changhoon Seo, Jong-Phil Kim, Jaeha Kim, Hyo-Sung Ahn, and Jeha Ryu. Robustly stable bilateral teleoperation under time-varying delays and data losses: an energy-bounding approach. *Journal of Mechanical Science and Technology*, 25 (8):2089, Sep 2011. ISSN 1976-3824. doi: 10.1007/s12206-011-0523-8. URL https://doi.org/10.1007/s12206-011-0523-8.

7 3GPP. Feasibility study on new services and markeets technology enablers for critical communications; stage 1. Technical report, 3GPP, October 2016. TR 22.862, Release 14.

8 3GPP. Study on new services and markets technology enablers. Technical report, 3GPP, September 2016. TR 22.891, Release 14.

9 T. B. Sheridan. Space teleoperation through time delay: review and prognosis. *IEEE Transactions on Robotics and Automation*, 9(5):592–606, Oct 1993. ISSN 1042-296X. doi: 10.1109/70.258052.

10 D. A. Lawrence. Stability and transparency in bilateral teleoperation. Technical Report 5, October 1993.

11 N. Pavon-Pulido, J.A. Lopez-Riquelme, J.J. Pinuaga-Cascales, J. Ferruz-Melero, and R. Morais Dos Santos. Cybi: A smart companion robot for elderly people: Improving teleoperation and telepresence skills by combining cloud computing technologies and fuzzy logic. In *Autonomous Robot Systems and Competitions (ICARSC), 2015 IEEE International Conference on*, pages 198–203, April 2015. doi: 10.1109/ICARSC.2015.40.

12 P. Arcara, C. Melchiorri. Control schemes for teleoperation with time delay: A comparative study. 38 (2002) 49–64, Robotics and Autonomous Systems, 2002.

13 P. Arcara. Control of Haptic and Robotic Telemanipulation Systems. Phd thesis, Università degli Studi di Bologna, 2014.

14 Jie Chen. Corrections to "On computing the maximal delay intervals for stability of linear delay systems". Technical Report 11, November 2000.

15 C. Melchiorri. Robotic Telemanipulation Systems: An Overview On Control Aspects. Technical report, International Federation of Automatic Control, 2003.

16 B. Hannaford. A design framework for teleoperators with kinesthetic feedback. *IEEE Transactions on Robotics and Automation*, 5(4):426–434, August 1989. ISSN 1042-296X. doi: 10.1109/70.88057.

17 William R. Ferrell. Delayed Force Feedback. *Human Factors*, 8(5):449–455, 1966. doi: 10.1177/001872086600800509. URL https://doi.org/10.1177/001872086600800509. PMID: 5966936.

18 W. S. Kim, B. Hannaford, and A. K. Fejczy. Force-reflection and shared compliant control in operating telemanipulators with time delay. Technical Report 2, April 1992.

19 W. S. Kim. Shared compliant control: a stability analysis and experiments. In *IEEE International Conference on Systems, Man, and Cybernetics*, pages 620–623, November 1990.

20 G. Niemeyer and J. J. E. Slotine. Stable adaptive teleoperation. *IEEE Journal of Oceanic Engineering*, 16(1):152–162, Jan 1991. ISSN 0364-9059. doi: 10.1109/48.64895.

21 R. J. Anderson and M. W. Spong. Bilateral control of teleoperators with time delay. *IEEE Transactions on Automatic Control*, 34(5):494–501, May 1989. ISSN 0018-9286. doi: 10.1109/9.24201.

22 G. J. Raju, G. C. Verghese, and T. B. Sheridan. Design issues in 2-port network models of bilateral remote manipulation. In *International Conference on Robotics and Automation*, pages 1316–1321 vol.3, May 1989.

23 Jonghyun Kim, Pyung Hun Chang, and Hyung-Soon Park. Transparent teleoperation using two-channel control architectures. In *IEEE/RSJ International Conference on Intelligent Robots and Systems*, pages 1953–1960, August 2005. doi: 10.1109/IROS.2005.1545404.

24 K. Hashtrudi-Zaad and S. E. Salcudean. Analysis and evaluation of stability and performance robustness for teleoperation control architectures. In *Proceedings 2000 ICRA. Millennium Conference. IEEE International Conference on Robotics and Automation. Symposia Proceedings (Cat. No.00CH37065)*, volume 4, pages 3107–3113 vol.4, 2000. doi: 10.1109/ROBOT.2000.845141.

25 K. Hastrudi-Zaad and S. E. Salcudean. On the use of local force feedback for transparent teleoperation. In *IEEE International Conference on Robotics and Automation*, volume 3, pages 1863–1869 vol.3, 1999. doi: 10.1109/ROBOT.1999.770380.

26 G. Niemeyer and J. J. E. Slotine. Towards force-reflecting teleoperation over the Internet. In *IEEE International Conference on Robotics and Automation*, volume 3, pages 1909–1915 vol.3, May 1998. doi: 10.1109/ROBOT.1998.680592.

27 Gunter Niemeyer and Jean-Jacques E. Slotine. Telemanipulation with Time Delays. *The International Journal of Robotics Research*, 23(9): 873–890, 2004. doi: 10.1177/0278364904045563.

28 L. Chan, F. Naghdy, and D. Stirling. Application of adaptive controllers in teleoperation systems: A survey. *IEEE Transactions on Human-Machine Systems*, 44(3):337–352, June 2014. ISSN 2168-2291. doi: 10.1109/THMS.2014.2303983.

29 Wen-Hong Zhu and S. E. Salcudean. Teleoperation with adaptive motion/force control. In *IEEE International Conference on Robotics and Automation*, volume 1, pages 231–237 vol.1, 1999. doi: 10.1109/ROBOT.1999.769976.

30 Wen-Hong Zhu and S. E. Salcudean. Stability guaranteed teleoperation: an adaptive motion/force control approach. *IEEE Transactions on Automatic Control*, 45(11):1951–1969, Nov 2000. ISSN 0018-9286. doi: 10.1109/9.887620.

31 S. Ganjefar, H. Momeni, F. Janabi Sharifi, and M. T. Hamidi Beheshti. Behavior of Smith predictor in teleoperation systems with modeling and delay time errors. In

IEEE Conference on Control Applications, volume 2, pages 1176–1180 vol.2, June 2003. doi: 10.1109/CCA.2003.1223177.

32 Andrew C. Smith and Keyvan Hashtrudi-Zaad. Smith Predictor Type Control Architectures for Time Delayed Teleoperation. *Internatonal Journal of Robotic Research*, 25(8):797–818, August 2006. ISSN 0278-3649. doi: 10.1177/0278364906068393. URL http://dx.doi.org/10.1177/0278364906068393.

33 G. Hirzinger, J. Heindl and K. Landzettel. Closer Control of Loops with Dead Time. Technical report, Chemistry Engineering Progress, Vol. 53, No. 5, pp. 217–219, 1957.

34 Z. Palmor. Stability properties of Smith dead-time compensator controllers. *International Journal of Control*, 32(6):937–949, 1980. doi: 10.1080/00207178008922900.

35 S. Munir and W. J. Book. Internet based teleoperation using wave variables with prediction. In *IEEE/ASME International Conference on Advanced Intelligent Mechatronics*, volume 1, pages 43–50 vol.1, 2001. doi: 10.1109/AIM.2001.936428.

36 Pietro Buttolo, Petter Braathen, and Blake Hannaford. Sliding Control of Force Reflecting Teleoperation: Preliminary Studies. *Presence: Teleoper. Virtual Environ.*, 3(2):158–172, January 1994. ISSN 1054-7460. doi: 10.1162/pres.1994.3.2.158. URL http://dx.doi.org/10.1162/pres.1994.3.2.158.

37 Jong Hyeon Park and Hyun Chul Cho. Sliding-mode controller for bilateral teleoperation with varying time delay. In *IEEE/ASME International Conference on Advanced Intelligent Mechatronics*, pages 311–316, September 1999. doi: 10.1109/AIM.1999.803184.

38 A. Hace and K. Jezernik. Bilateral teleoperation by sliding mode control and reaction force observer. In *IEEE International Symposium on Industrial Electronics*, pages 1809–1816, July 2010. doi: 10.1109/ISIE.2010.5637717.

39 Aleš Hace and Marko Franc. Bilateral teleoperation by sliding mode control and reaction force observer. Technical report, IEEE TRANSACTIONS ON INDUSTRIAL INFORMATICS, VOL.9, NO.3, 2013.

40 B. Hannaford and Jee-Hwan Ryu. Time-domain passivity control of haptic interfaces. *Robotics and Automation, IEEE Transactions on*, 18(1):1–10, Feb 2002. ISSN 1042-296X. doi: 10.1109/70.988969.

41 Jee-Hwan Ryu, Dong-Soo Kwon, and B. Hannaford. Stable teleoperation with time-domain passivity control. *Robotics and Automation, IEEE Transactions on*, 20(2):365–373, April 2004. ISSN 1042-296X. doi: 10.1109/TRA.2004.824689.

42 Jee-Hwan Ryu, Jordi Artigas, and Carsten Preusche. A passive bilateral control scheme for a teleoperator with time-varying communication delay. *Mechatronics*, 20(7):812 –823, 2010. ISSN 0957-4158. doi: http://dx.doi.org/10.1016/j.mechatronics.2010.07.006. URL http://www.sciencedirect.com/science/article/pii/S0957415810001303. Special Issue on Design and Control Methodologies in Telerobotics.

43 Hongbing Li, Kotaro Tadano, and Kenji Kawashima. Model-based passive bilateral teleoperation with time delay. *Transactions of the Institute of Measurement and Control*, 36(8):1010–1023, 2014. doi: 10.1177/0142331214530423. URL http://tim.sagepub.com/content/36/8/1010.abstract.

44 Jong-Phil Kim and Jeha Ryu. Robustly stable haptic interaction control using an energy-bounding algorithm. *The International Journal of Robotics Research*,

29(6):666–679, 2010. doi: 10.1177/0278364909338770. URL http://ijr.sagepub.com/content/29/6/666.abstract.

45 L. Marton, J.A. Esclusa, P. Haller, and T. Vajda. Passive bilateral teleoperation with bounded control signals. In *Industrial Informatics (INDIN), 2013 11th IEEE International Conference on*, pages 337–342, July 2013. doi: 10.1109/INDIN.2013.6622906.

46 Yongjun Hou and G.R. Luecke. Time delayed teleoperation system control, a passivity-based method. In *Advanced Robotics, 2005. ICAR '05. Proceedings, 12th International Conference on*, pages 796–802, July 2005. doi: 10.1109/ICAR.2005.1507499.

47 Xiao Xu, Burak Cizmeci, Clemens Schuwerk, and Eckehard Steinbach. Haptic data reduction for time-delayed teleoperation using the time domain passivity approach. In *IEEE World Haptics Conference, Chicago, USA*, Jun 2015a.

48 J. Rebelo and A. Schiele. Time domain passivity controller for 4-channel time-delay bilateral teleoperation. *Haptics, IEEE Transactions on*, 8(1): 79–89, Jan 2015. ISSN 1939-1412. doi: 10.1109/TOH.2014.2363466.

49 Jordi Artigas, Jee-Hwan Ryu, C. Preusche, and G. Hirzinger. Network representation and passivity of delayed teleoperation systems. In *Intelligent Robots and Systems (IROS), 2011 IEEE/RSJ International Conference on*, pages 177–183, Sept 2011. doi: 10.1109/IROS.2011.6094919.

50 Jee-Hwan Ryu, Dong-Soo Kwon, and B. Hannaford. Stability guaranteed control: time domain passivity approach. *Control Systems Technology, IEEE Transactions on*, 12(6):860–868, Nov 2004. ISSN 1063-6536. doi: 10.1109/TCST.2004.833648.

51 Probal Mitra and Günter Niemeyer. Model-mediated telemanipulation. *The International Journal of Robotics Research*, 27(2):253–262, 2008. doi: 10.1177/0278364907084590. URL http://ijr.sagepub.com/content/27/2/253 .abstract.

52 Carolina Passenberg, Angelika Peer, and Martin Buss. A survey of environment-, operator-, and task-adapted controllers for teleoperation systems. *Mechatronics*, 20(7):787–801, 2010a. ISSN 0957-4158. doi: http://dx.doi.org/10.1016/j .mechatronics.2010.04.005. Special Issue on Design and Control Methodologies in Telerobotics.

53 C. Passenberg, A. Peer, and M. Buss. Model-mediated teleoperation for multi-operator multi-robot systems. In *Intelligent Robots and Systems (IROS), 2010 IEEE/RSJ International Conference on*, pages 4263–4268, Oct 2010b. doi: 10.1109/IROS.2010.5653012.

54 Bert Willaert, Hendrik Van Brussel, and Günter Niemeyer. Stability of model-mediated teleoperation: Discussion and experiments. In Poika Isokoski and Jukka Springare, editors, *Haptics: Perception, Devices, Mobility, and Communication*, volume 7282 of *Lecture Notes in Computer Science*, pages 625–636. Springer Berlin Heidelberg, 2012. ISBN 978-3-642-31400-1. doi: 10.1007/978-3-642-31401-8_55.

55 Xiao Xu, C. Schuwerk, and E. Steinbach. Passivity-based model updating for model-mediated teleoperation. In *Multimedia Expo Workshops (ICMEW), 2015 IEEE International Conference on*, pages 1–6, June 2015. doi: 10.1109/ICMEW.2015.7169831.

56 F.T. Buzan and T.B. Sheridan. A model-based predictive operator aid for telemanipulators with time delay. In *Systems, Man and Cybernetics, 1989. Conference
   Proceedings, IEEE International Conference on*, pages 138–143 vol.1, Nov 1989.
   doi: 10.1109/ICSMC.1989.71268.

57 P.A. Prekopiou, S.G. Tzafestas, and W.S. Harwin. Towards
   variable-time-delays-robust telemanipulation through master state prediction.
   In *Advanced Intelligent Mechatronics, 1999. Proceedings. 1999 IEEE/ASME
   International Conference on*, pages 305–310, 1999. doi: 10.1109/AIM.1999.803183.

58 Ismal Belghit France. Predictive algorithms for distant touching, 2002.

59 C. Weber, V. Nitsch, U. Unterhinninghofen, B. Farber, and M. Buss. Position and
   force augmentation in a telepresence system and their effects on perceived realism.
   In *EuroHaptics conference, 2009 and Symposium on Haptic Interfaces for Virtual
   Environment and Teleoperator Systems. World Haptics 2009. Third Joint*, pages
   226–231, March 2009. doi: 10.1109/WHC.2009.4810803.

60 D. Verscheure, J. Swevers, H. Bruyninckx, and J. De Schutter. On-line identification
   of contact dynamics in the presence of geometric uncertainties. In *Robotics and
   Automation, 2008. ICRA 2008. IEEE International Conference on*, pages 851–856,
   May 2008. doi: 10.1109/ROBOT.2008.4543311.

61 J.E. Colgate and J.M. Brown. Factors affecting the z-width of a haptic display. In
   *Robotics and Automation, 1994. Proceedings, 1994 IEEE International Conference
   on*, pages 3205–3210 vol.4, May 1994. doi: 10.1109/ROBOT.1994.351077.

62 Hong Tan, J. Radcliffe, Book No. Ga, Hong Z. Tan, Brian Eberman, Mandayam A.
   Srinivasan, and Belinda Cheng. Human factors for the design of force-reflecting
   haptic interfaces, 1994.

63 T.L. Brooks. Telerobotic response requirements. In *Systems, Man and Cybernetics,
   1990. Conference Proceedings, IEEE International Conference on*, pages 113–120,
   Nov 1990. doi: 10.1109/ICSMC.1990.142071.

64 K.B. Shimoga. A survey of perceptual feedback issues in dexterous telemanipulation. i. finger force feedback. In *Virtual Reality Annual International Symposium,
   1993, 1993 IEEE*, pages 263–270, Sep 1993. doi: 10.1109/VRAIS.1993.380770.

65 K.B. Shimoga. A survey of perceptual feedback issues in dexterous telemanipulation. ii. finger touch feedback. In *Virtual Reality Annual International Symposium,
   1993, 1993 IEEE*, pages 271–279, Sep 1993. doi: 10.1109/VRAIS.1993.380769.

66 P.G. Otanez, James R. Moyne, and D.M. Tilbury. Using deadbands to reduce
   communication in networked control systems. In *American Control Conference, 2002. Proceedings of the 2002*, volume 4, pages 3015–3020 vol.4, 2002.
   doi: 10.1109/ACC.2002.1025251.

67 P. Hinterseer, S. Hirche, S. Chaudhuri, E. Steinbach, and M. Buss. Perception-based
   data reduction and transmission of haptic data in telepresence and teleaction systems. *IEEE Transactions on Signal Processing*, 56(2):588–597, February 2008. ISSN
   1053-587X. doi: 10.1109/TSP.2007.906746.

68 Q. Nasir and E. Khalil. Perception based adaptive haptic communication
   protocol (pahcp). In *Computer Systems and Industrial Informatics (ICC-
   SII), 2012 International Conference on*, pages 1–6, Dec 2012. doi: 10.1109/
   ICCSII.2012.6454426.

69 C. Shahabi, A. Ortega, and M.R. Kolahdouzan. A comparison of different haptic
   compression techniques. In *Multimedia and Expo, 2002. ICME '02. Proceedings.*

2002 IEEE International Conference on, volume 1, pages 657–660 vol.1, 2002. doi: 10.1109/ICME.2002.1035867.

70 Yonghee You and Mee Young Sung. Haptic data transmission based on the prediction and compression. In *Communications, 2008. ICC '08. IEEE International Conference on*, pages 1824–1828, May 2008. doi: 10.1109/ICC.2008.350.

71 H. Tanaka and K. Ohnishi. Haptic data compression/decompression using dct for motion copy system. In *Mechatronics, 2009. ICM 2009. IEEE International Conference on*, pages 1–6, April 2009. doi: 10.1109/ICMECH.2009.4957156.

72 Ahmet Kuzu, Eray A Baran, Seta Bogosyan, Metin Gokasan, andAsif Sabanovic. Wavelet packet transform-based compression for teleoperation. *Proceedings of the Institution of Mechanical Engineers, Part I: Journal of Systems and Control Engineering*, 229(7):639–651, 2015. doi: 10.1177/0959651815575438. URL http://pii.sagepub.com/content/229/7/639.abstract.

73 E. Steinbach, S. Hirche, J. Kammerl, I. Vittorias, and R. Chaudhari. Haptic data compression and communication. *Signal Processing Magazine, IEEE*, 28(1):87–96, Jan 2011. ISSN 1053-5888.

74 V. Nitsch, B. Farber, L. Geiger, P. Hinterseer, and E. Steinbach. An experimental study of lossy compression in a real telepresence and teleaction system. In *Haptic Audio visual Environments and Games, 2008. HAVE 2008. IEEE International Workshop on*, pages 75–80, Oct 2008. doi: 10.1109/HAVE.2008.4685302.

75 E.H. Weber. *De Pulsu, resorptione, auditu et tactu: Annotationes anatomicae et physiologicae … Annotationes anatomicae et physiologicae*. C.F. Koehler, 1834. URL https://books.google.co.uk/books?id=bdI-AAAAYAAJ.

76 Jalal Awed, ImadH. Elhajj, and Nadiya Slobodenyuk. Haptic force perception in bimanual manipulation. In Poika Isokoski and Jukka Springare, editors, *Haptics: Perception, Devices, Mobility, and Communication*, volume 7283 of *Lecture Notes in Computer Science*, pages 1–6. Springer Berlin Heidelberg, 2012. ISBN 978-3-642-31403-2. doi: 10.1007/978-3-642-31404-9_1.

77 S. Allin, Y. Matsuoka, and R. Klatzky. Measuring just noticeable differences for haptic force feedback: implications for rehabilitation. In *Haptic Interfaces for Virtual Environment and Teleoperator Systems, 2002. HAPTICS 2002. Proceedings. 10th Symposium on*, pages 299–302, 2002. doi: 10.1109/HAPTIC.2002.998972.

78 P. Hinterseer, E. Steinbach, S. Hirche, and M. Buss. A novel, psychophysically motivated transmission approach for haptic data streams in telepresence and teleaction systems. In *Acoustics, Speech, and Signal Processing, 2005. Proceedings. (ICASSP '05). IEEE International Conference on*, volume 2, pages ii/1097–ii/1100 Vol. 2, March 2005.

79 N. Sakr, Jilin Zhou, Nicolas D. Georganas, Jiying Zhao, and E.M. Petriu. Robust perception-based data reduction and transmission in telehaptic systems. In *EuroHaptics conference, 2009 and Symposium on Haptic Interfaces for Virtual Environment and Teleoperator Systems. World Haptics 2009. Third Joint*, pages 214–219, March 2009. doi: 10.1109/WHC.2009.4810839.

80 A. Bhardwaj, O. Dabeer, and S. Chaudhuri. Can we improve over weber sampling of haptic signals? In *Information Theory and Applications Workshop (ITA), 2013*, pages 1–6, Feb 2013. doi: 10.1109/ITA.2013.6502929.

81 Amit Bhardwaj, Subhasis Chaudhuri, and Onkar Dabeer. Design and analysis of predictive sampling of haptic signals. *ACM Trans. Appl. Percept.*, 11(4):16:1–16:20,

December 2014. ISSN 1544-3558. doi: 10.1145/2670533. URL http://doi.acm.org/10.1145/2670533.

82 P. Hinterseer, E. Steibach, and S. Chaudhuri. Model based data compression for 3d virtual haptic teleinteraction. In *Consumer Electronics, 2006. ICCE '06. 2006 Digest of Technical Papers. International Conference on*, pages 23–24, Jan 2006. doi: 10.1109/ICCE.2006.1598291.

83 P. Hinterseer, E. Steinbach, and S. Chaudhuri. Perception-based compression of haptic data streams using kalman filters. In *Acoustics, Speech and Signal Processing, 2006. ICASSP 2006 Proceedings. 2006 IEEE International Conference on*, volume 5, pages V–V, May 2006. doi: 10.1109/ICASSP.2006.1661315.

84 N. Sakr, N.D. Georganas, J. Zhao, and X. Shen. Motion and force prediction in haptic media. In *Multimedia and Expo, 2007 IEEE International Conference on*, pages 2242–2245, July 2007. doi: 10.1109/ICME.2007.4285132.

85 Fenghua Guo, Caiming Zhang, and Yan He. Haptic data compression based on a linear prediction model and quadratic curve reconstruction. *Journal of Software*, 9(11), 2014. URL http://ojs.academypublisher.com/index.php/jsw/article/view/jsw091127962803.

86 Xiao Xu, J. Kammerl, R. Chaudhari, and E. Steinbach. Hybrid signal-based and geometry-based prediction for haptic data reduction. In *Haptic Audio Visual Environments and Games (HAVE), 2011 IEEE International Workshop on*, pages 68–73, Oct 2011. doi: 10.1109/HAVE.2011.6088394.

87 S. Payandeh and Jae-Young Lee. User-centered force signal processing for internet-based telemanipulation: An overview. In *Systems, Man and Cybernetics (SMC), 2014 IEEE International Conference on*, pages 3978–3983, Oct 2014. doi: 10.1109/SMC.2014.6974553.

88 Katherine J. Kuchenbecker, Jonathan Fiene, and Gunter Niemeyer. Improving contact realism through event-based haptic feedback. *IEEE Transactions on Visualization and Computer Graphics*, 12(2):219–230, March 2006. ISSN 1077-2626. doi: 10.1109/TVCG.2006.32. URL http://dx.doi.org/10.1109/TVCG.2006.32.

89 W. McMahan, J.M. Romano, A.M. Abdul Rahuman, and K.J. Kuchenbecker. High frequency acceleration feedback significantly increases the realism of haptically rendered textured surfaces. In *Haptics Symposium, 2010 IEEE*, pages 141–148, March 2010. doi: 10.1109/HAPTIC.2010.5444665.

90 A.M. Okamura, J.T. Dennerlein, and R.D. Howe. Vibration feedback models for virtual environments. In *Robotics and Automation, 1998. Proceedings. 1998 IEEE International Conference on*, volume 1, pages 674–679 vol.1, May 1998. doi: 10.1109/ROBOT.1998.677050.

91 Katherine J. Kuchenbecker, Joseph M. Romano, and William McMahan. Haptography: Capturing and recreating the rich feel of real surfaces. In *Int. Symp. on Robotics Research (invited paper)*, volume 70 of *Springer Tracts in Advanced Robotics*, pages 245–260. Springer, 2009. ISBN 978-3-642-19456-6.

92 V.L. Guruswamy, J. Lang, and Won-Sook Lee. Iir filter models of haptic vibration textures. *Instrumentation and Measurement, IEEE Transactions on*, 60(1):93–103, Jan 2011. ISSN 0018-9456. doi: 10.1109/TIM.2010.2065751.

93 S. Okamoto and Y. Yamada. Perceptual properties of vibrotactile material texture: Effects of amplitude changes and stimuli beneath detection thresholds. In *System*

Integration (SII), 2010 IEEE/SICE International Symposium on, pages 384–389, Dec 2010. doi: 10.1109/SII.2010.5708356.

94 James C. Craig. Difference threshold for intensity of tactile stimuli. *Perception & Psychophysics*, 11(2):150–152, 1972. ISSN 0031-5117. doi: 10.3758/BF03210362. URL http://dx.doi.org/10.3758/BF03210362.

95 Helena Pongrac. Vibrotactile perception: examining the coding of vibrations and the just noticeable difference under various conditions. *Multimedia Systems*, 13 (4):297–307, 2008. ISSN 0942-4962. doi: 10.1007/s00530-007-0105-x. URL http://dx .doi.org/10.1007/s00530-007-0105-x.

96 Christian Hatzfeld and Roland Werthschützky. Just noticeable differences of low-intensity vibrotactile forces at the fingertip. In Poika Isokoski and Jukka Springare, editors, *Haptics: Perception, Devices, Mobility, and Communication*, volume 7283 of *Lecture Notes in Computer Science*, pages 43–48. Springer Berlin Heidelberg, 2012. ISBN 978-3-642-31403-2. doi: 10.1007/978-3-642-31404-9_8.

97 Rahul Chaudhari, Burak Çizmeci, Katherine J. Kuchenbecker, Seungmoon Choi, and Eckehard Steinbach. Low bitrate source-filter model based compression of vibrotactile texture signals in haptic teleoperation. In *Proceedings of the 20th ACM International Conference on Multimedia*, MM '12, pages 409–418, New York, NY, USA, 2012. ACM. ISBN 978-1-4503-1089-5. doi: 10.1145/2393347.2393407. URL http://doi.acm.org/10.1145/2393347.2393407.

98 International Telecommunication Union and International Telecommunication Union Telecommunication Standardization Sector. *Coding of Speech at 8 Kbit/s Using Conjugate-structure Algebraic-code-excited Linear-prediction (CS-ACELP): Reduced complexity 8 kbit/s CS-ACELP speech codec. Annex A*. ITU-T recommendation. International Telecommunication Union, 1996. URL https://www.itu.int/rec/T-REC-G.729/e.

99 R. Chaudhari, C. Schuwerk, M. Danaei, and E. Steinbach. Perceptual and bitrate-scalable coding of haptic surface texture signals. *Selected Topics in Signal Processing, IEEE Journal of*, 9(3):462–473, April 2015. ISSN 1932-4553. doi: 10.1109/JSTSP.2014.2374574.

100 E. Zwicker and R. Feldtkeller. *Das Ohr als Nachrichtenempfänger*. Monographien der elektrischen Nachrichtentechnik. Hirzel, 1967. ISBN 9783777601045. URL https://books.google.co.uk/books?id=kyZtQgAACAAJ.

101 A. Tatematsu, Y. Ishibashi, N. Fukushima, and S. Sugawara. Qoe assessment in tele-operation with 3d video and haptic media. In *2011 IEEE International Conference on Multimedia and Expo*, pages 1–6, July 2011. doi: 10.1109/ICME.2011.6012159.

102 Xun Liu, M. Dohler, T. Mahmoodi, and Hongbin Liu. Challenges and opportunities for designing tactile codecs from audio codecs. In *2017 European Conference on Networks and Communications (EuCNC)*, pages 1–5, June 2017.

103 Sandra Hirche and Martin Buss. Transparent data reduction in networked telepresence and teleaction systems. part ii: Time-delayed communication. *Presence: Teleoperators and Virtual Environments*, 16 (5):532–542, Sep 2007. ISSN 1054-7460. doi: 10.1162/pres.16.5.532. URL http://dx.doi.org/10.1162/pres.16.5.532.

104 I. Vittorias, J. Kammerl, S. Hirche, and E. Steinbach. Perceptual coding of haptic data in time-delayed teleoperation. In *EuroHaptics conference, 2009 and Symposium on Haptic Interfaces for Virtual Environment and Teleoperator*

Systems. World Haptics 2009. Third Joint, pages 208–213, March 2009. doi: 10.1109/WHC.2009.4810811.

105 Xiao Xu, B. Cizmeci, A. Al-Nuaimi, and E. Steinbach. Point cloud-based model-mediated teleoperation with dynamic and perception-based model updating. *Instrumentation and Measurement, IEEE Transactions on*, 63(11):2558–2569, Nov 2014. ISSN 0018-9456. doi: 10.1109/TIM.2014.2323139.

4

5G-Enhanced Smart Grid Services

Muhammad Ismail, Islam Safak Bayram, Khalid Qaraqe and Erchin Serpedin

The advanced services that will be offered by future smart grids call for a tight integration between power systems and communication networks. To enable enhanced power grid monitoring, effective demand side management, and appropriate charging and discharging coordination of electric vehicles, huge amounts of data, with stringent quality-of-service requirements, need to be transferred among different entities of the grid in real time. The currently existing networks are incapable of handling this volume of traffic in a flexible, scalable, and dynamic way. Fortunately, the novel concepts of 5G networks such as network slicing, cloud computing, software defined networking, and network function virtualization offer several attractive features that perfectly fit the communication requirements of the smart grid services. Through this chapter, we aim to shed the light on the 5G key concepts and how they can be extremely beneficial in supporting the advanced smart grid services. This chapter first introduces the smart grid environment and discusses some of the future services that will be supported in the future smart grids. These services are broadly classified into two categories, namely data collection and management services that target enhanced grid monitoring capabilities, and control and operation services that deal with demand side management and electric vehicle charging and discharging coordination. Then we illustrate how the 5G novel concepts of cloud computing, software defined networking, and network function virtualization can be employed to provide flexible, scalable, and secure services in the smart grid. Future research directions are finally discussed to deal with the open challenging issues.

4.1 Introduction

The past decade has witnessed a revolutionary transformation in the power grid infrastructure, management, and operation to enable efficient, reliable, sustainable, clean, and secure electric grid. Traditionally, the power grid followed a centralized generation scheme, where electricity is generated in distant large power plants (bulk generators) and then carried for several kilometers through a transmission grid to be delivered to the end customers via the distribution grid. Such a traditional system relies on few sensors that are distributed across the grid to carry some information back to the grid operator

Enabling 5G Communication Systems to Support Vertical Industries, First Edition.
Edited by Muhammad Ali Imran, Yusuf Abdulrahman Sambo and Qammer H. Abbasi.
© 2019 John Wiley & Sons Ltd. Published 2019 by John Wiley & Sons Ltd.

in order to take limited control actions. As a result, the traditional power grid depends on manual monitoring and restoration mechanisms, and hence may suffer from occasional failures and blackouts. The introduction of modern concepts and the integration of the power grid with advanced communication networks paved the way for a smart power grid that promises an enhanced electric grid operation.

Distributed generation is a key concept that has been introduced in the smart grid. Through this concept, power generation is enabled in the distribution grid via small-scale distributed generators (DGs). These DGs can be dispatchable units that run on non-renewable fuel (e.g., diesel and gas) or renewable DG units such as solar panels and wind turbines. Hence the customers can generate their own electricity at their premises. As a result of installing such DG units within the distribution grid, a new paradigm has been introduced in the smart grid, namely the microgrid. The smart grid can be regarded as a collection of microgrids. Each microgrid presents a set of DG units, energy storage systems (ESSs), and loads. Electricity now can flow either from the bulk generators to the microgrid or can be sold from the customers in the microgrid back to the electric utility company. Hence the smart grid enables two-way flow of electricity. To ensure more efficient and reliable operation of the smart grid, a massive number of sensors are deployed throughout the power grid, which presents better observability of the grid, and as a result, pervasive control actions can be taken. These control actions can be transmitted autonomously from the grid control center in a timely manner to eliminate grid failures and blackouts and present a more resilient operation of the grid via advanced self-monitoring and self-healing mechanisms. Hence, unlike the traditional grid, the smart grid relies on two-way flow of information and electricity to enhance the overall performance of the electricity sector. Thanks to the two-way flow of information, new concepts such as demand-side management (DSM) can be introduced in the smart grid. Specifically, through price control or direct load control (DLC), the grid operator can reduce the peak power demand in the grid, which is known as peak shaving. As a result, in the smart grid, the load will follow the available generation rather than the traditional approach where the generation used to follow the load. Hence, expensive grid expansion plans can be deferred, which saves future investments. Furthermore, due to the enhanced control mechanisms, new electric loads, i.e., electric vehicles (EVs) and ESS units, can be introduced to the grid with minimal impact. This is possible because of novel charging and discharging coordination mechanisms.

It is evident from the above discussion that many of the aforementioned novel functionalities of the smart grid are possible thanks to the tight integration between the power grid and the communication networks. In general, there exist two types of smart grid services that require communication network support, as follows: 1) data collection and management service, which is concerned with collecting the readings from the massive number of sensors that are deployed across the grid and the meters within the customers' premises and exchanging this data among different entities of the grid for decision making and 2) control and operation service, which is responsible for reaching optimal control actions either for emergency mitigation or for ensuring efficient normal operation of the grid, as in DSM and EV charging and discharging coordination decisions. Given the nature of these two services, the following key requirements must be available in communication networks in order to provide the necessary support: 1) massive access: this is essential in order to be able to collect the readings from all the sensors and meters deployed within the smart grid. These connections are clearly not human-to-human connections, but mainly machine-to-machine and machine-to-human communications. It should be highlighted that such a massive access for metering data collection should be done with very high reliability, i.e., to ensure a packet loss ratio (PLR) below 1%, and 2) ultra-low latency: this is necessary for ensuring that emergency control actions are taken at a proper time. Unfortunately, the 4G networks are unable to provide such low latency even at light load conditions. Specifically, experimental results have proven that the 4G LTE radio networks can achieve 15–20 milliseconds in the uplink and as low as 4 milliseconds in the downlink, for a PLR of 10^{-1} [1]. To achieve an improved PLR of 10^{-5}, the latency has to be increased to 40–60 milliseconds. In addition, the 4G networks are not designed to support such massive machine-type connections, also with high reliability. Furthermore, in 4G networks, all the aforementioned smart grid services will be supported on the same network along with other type of services, e.g., human-to-human communications. The 4G networks present the same support and network functions to all kind of services regardless of their requirements. In turn, this does not meet the service isolation and differentiation that should be available in the network supporting the smart grid services.

On the other hand, the key concepts that are introduced in the 5G networks present a perfect match to the communication requirements of the smart grid services. Among the three service scenarios defined in the 5G networks, two service scenarios fit very well the smart grid service requirements. Specifically, the massive machine type communication (mMTC) and the ultra-reliability low-latency communication (URLLC) scenarios present a perfect fit for the communication requirements of the data collection and control services of the smart grid, respectively. The mMTC service supports a massive access of machines, namely, up to 10 million connections/m^2 [2]. In addition, the URLLC service supports applications with stringent latency requirements down to 1 millisecond with extremely high reliability [2]. More importantly, the concept of network slicing that is introduced in 5G networks is able to construct different (virtual) network entities that are logically isolated and hence can support different applications on the same network without interfering with each other. As a result, two network slices for mMTC and URLLC can be dedicated to support data collection and control services in the smart grid, while the same physical network can also be supporting other applications and industries. Furthermore, the 5G key concepts of network function virtualization (NFV), software-defined networking (SDN), and cloud computing offer a more efficient and flexible way for handling the smart grid big data and reaching optimal operational decisions in a timely manner.

The objective of this chapter is to shed the light on the 5G key concepts and how they can be extremely beneficial in supporting the smart grid services for data collection and control services. Towards this objective, we first present an overview of smart grid fundamentals with a special focus on data collection via smart meters (SMs) and phasor measurement units (PMUs) and the operation decision making for DSM and EV charging and discharging coordination. Then, we highlight the 5G key concepts in terms of network slicing, NFV, SDN, and cloud computing. Based on these foundations, the rest of the chapter discusses a set of application scenarios where the 5G key concepts play a vital role in supporting the smart grid data collection and control services. Finally, future research directions are defined.

4.2 Smart Grid Services and Communication Requirements

This section presents an overview of some of the key concepts in the smart grid and the communication requirements of different smart grid services. A brief summary is also presented for sake of completeness to discuss 5G fundamental concepts that help in addressing the communication requirements of the smart grid services.

4.2.1 Smart Grid Fundamentals

This subsection presents an overview of data collection and control services in smart grids. Specifically, we briefly discuss some issues related to SMs and PMUs for data collection in smart grids. Also, we present the key principles related to control services in smart grid for DSM and EV charging and discharging coordination.

4.2.1.1 Data Collection and Management Services

The advanced functionalities to be offered by the smart grid are possible thanks to the sophisticated monitoring system that is expected to exist in the smart grid. Such a system monitors various components within the grid and provides continuous measurements from these components to reflect their current status, which presents sufficient information for the grid operator to take necessary optimal control actions in a timely manner. In providing this information, the monitoring system relies on a massive number of SMs and PMUs that are installed at various locations in the power grid [3].

The first location where such meters are needed is at the generation units, in order to monitor the electricity generation in terms of the amount of injected power and frequency. This is necessary to measure the power quality throughout the grid. Also, this helps in ensuring the balance between both supply and demand. These meters will be of vital importance at the DG units such as wind turbines and solar panels that are installed at the customers' premises to guarantee voltage and frequency stability within the power grid. Furthermore, these meters are needed for billing purposes when a customer sells the electricity generated at his/her DG unit back to the grid. The second location where meters are needed is at the ESS units in order to measure the batteries' state-of-charge (SOC). With the presence of many renewable-based DG units, and given the intermittent nature of such units, a frequent update of the SOC of the ESS units within the microgrid is needed so as to guarantee a balance between supply and demand.

In addition, SMs and sensors/actuators with two-way communication capabilities should be widely distributed across the transmission and distribution networks of the smart grid. Such devices measure the power flow within the buses and monitor substations, transformers, capacitor banks, etc. Using the power flow measurements across the grid, a state estimation process takes place at the smart grid control unit to determine the voltage phase angles at the buses (assuming a unity magnitude in a DC state estimation process). On the other hand, advanced PMUs can directly measure the voltage phase angles at different buses and these readings are synchronized using the global positioning system (GPS) radio clock [4]. In general, given the voltage values at different buses, voltage stability can be guaranteed within the grid and over-voltage and under-voltage situations can be alleviated.

The last location where SMs are needed is at the customers' premises in order to monitor the energy consumption. Such SMs are part of the advanced metering infrastructure (AMI), which includes SMs, communication networks, and meter data management systems (MDMS). Home SMs are able to provide precise readings that are recorded/displayed at the meter and also reported to the electricity utility company every 5, 15, or 30 minutes. This is extremely beneficial for the customers to monitor their energy consumption and as a result control their consumption pattern for a reduced electricity bill. Furthermore, this information is used by the electric utility company for automatic billing. An AMI testbed for monitoring of energy consumption in a residential unit is shown in Figure 4.1. A 3-phase Smappee energy monitoring device is installed on the main circuit panel of the house. Clamp meters are utilized to measure the current flow. To power the monitoring device, a plug-in power supply is used and the Smappee device determines the voltage drop across the power supply. Using the voltage and current parameters, the energy consumption data of the house is calculated every 5 minutes. The Smappee device is connected to the internet via WiFi. The energy consumption data of the house can be accessed through the Smappee cloud for real-time monitoring and archival records are also saved at a local server. In this testbed, the Smappee device, WiFi connection, and Smappee cloud and local server represent the SM, communication network, and MDMS of the AMI. Figure 4.2 shows the monthly average energy consumption data that is monitored by the AMI testbed in the period from July to November 2017. It can be noticed from the results that there is a significant difference in the energy consumption between summer months such as July and August and winter months such as November. Such a difference is mainly due to the heavy dependence on air-conditioning during the hot summer months. This difference is almost up to 6 kW power consumption, which opens the room for a great energy saving if appropriate DSM program is adopted during the summer months while

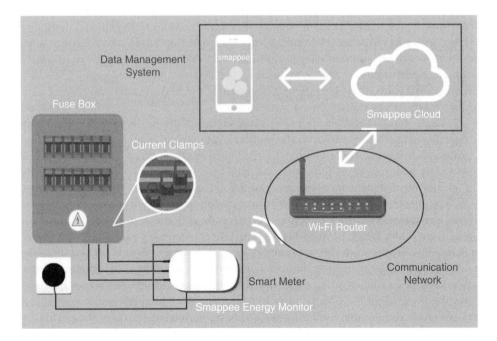

Figure 4.1 Illustration of an AMI testbed [5].

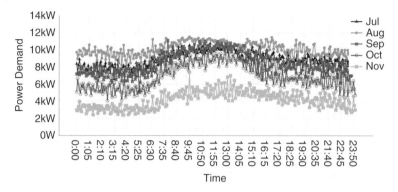

Figure 4.2 Collected energy consumption data from a residential unit using the AMI testbed [5].

considering the customers' comfort level. The two-way communication capabilities of these SMs enable active interaction between the customers and the electricity utility company. As a result, peak load reduction techniques can be implemented, for instance via real-time pricing mechanisms, which is extremely beneficial for the grid operator in order to defer unnecessary grid expansion plans. In this case, when a peak load is expected by the grid operator, a high electricity price signal is sent to the SM, and as a result the customer is encouraged to reduce his/her electricity consumption for a reduced electricity bill. In addition, advanced SMs are also able to control the operation of smart appliances within the customer premises. Hence, an effective DLC program can be implemented where a control signal is sent from the grid operator to the customer's SM in order to adjust or defer the operation of a device or to switch it off as a means to peak load shaving in exchange for some monetary incentives. This kind of interaction is part of the smart home paradigm. Furthermore, thanks to the continuous monitoring of the customers' energy consumption, SMs provide an effective means to detect electricity thefts.

Deployment plans of SMs in different countries reveal that a massive number of such devices are either already installed or will be installed in the near future [6]. For instance, in USA, over 36 million SMs are already installed since 2010. In China, 230 million SMs are installed since 2015. Moreover, 150 million SMs are planned to be installed in India by 2025. With the regular reporting of such meter readings, massive connections are expected. In addition, these SMs open the door to several security and privacy threats [7]. Specifically, frequent measurements of the customers' power consumptions (e.g., every 5 minutes) enable eavesdroppers to analyze the customers' habits and hence provide information on whether a customer is currently at home or not. This information can be misused by burglars for instance. In addition, since SMs can control the smart appliances within the customer's house for DSM, an attacker might hack into the SMs, and hence control such smart appliances. Furthermore, electricity theft attacks can now be launched in a cyber manner, where a malicious user hacks into the SM and manipulates the integrity of the SM readings in order to reduce his/her electricity bill. The massive access requirement and the associated data management issues along with the security and privacy challenges of SMs, sensors, and actuators are well addressed by the 5G communication networks, as will be discussed shortly.

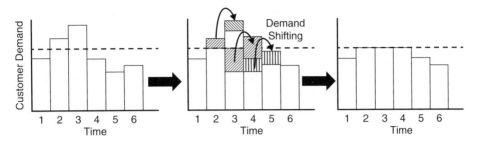

Figure 4.3 Illustration of peak load shifting in DSM.

4.2.1.2 Control and Operation Services

In this subsection, we focus on two control and operation services in smart grid, namely DSM and EV charging and discharging coordination.

Demand Side Management

DSM refers to a set of methods and strategies that are applied by the grid operator in order to modify the customers' energy demands. Such an approach is adopted to reduce the peak load within the grid, and therefore cope with the limited energy production, as shown in Figure 4.3. Consequently, in the future smart grid, energy demands will follow the generation, rather than the traditional approach where the generation used to follow the energy demands. This is possible due to the two-way communication within the smart grid that promotes customer-utility interaction and enables more efficient energy usage.

DSM presents several benefits, including economic advantages, reduced greenhouse gas (GHG) emissions, and improved grid resiliency and power quality [8]. Specifically, both the utilities and the customers benefit from the economic advantages of DSM. First, the customers enjoy monetary incentives and differentiated tariffs for shifting their peak hour demand. Furthermore, by reducing the peak demand in the grid, the required upgrades to cater for the increased demand will be postponed and may now occur gradually over time, which in turn saves future investments. In addition, DSM offers a great opportunity to cut the GHG emissions since most of the pollutant electricity generation units are employed during the peak load hours, which can be alleviated using effective DSM strategies. Moreover, intelligent DSM techniques will help in reducing the mean service interruption duration, which will further protect the loads against short-term effects such as voltage spikes, dips, and surges. Consequently, both grid resiliency and power quality will be improved.

In order to develop a mathematical formulation for the DSM strategies, mathematical models of electric loads are required. In general, electric loads convert electricity into another form of energy that is deemed useful for the customers. Based on their physical properties, electric loads can be classified into resistive, inductive, capacitive, non-linear, and composite types [8]. Resistive loads convert electric energy into heat, as in incandescent lights, ovens, toasters, and heaters. Such resistive loads usually follow on-off models, where the load draws fixed power during the on state and zero or minimal amount of power is drawn during the off state. On the other hand, inductive loads convert electric energy into motion using AC motors, as in refrigerators, vacuum cleaners, and air conditioners. These loads follow on-off decay models, where the power drawn during the on state reaches a stable level after a decay rate μ. Non-linear loads such as TV sets, computers, and fluorescent lights draw current that does not follow a sinusoidal pattern. Composite loads include more than one type of electric properties, as in the refrigerator, which can be a composite of inductive load (for the compressor) and resistive load (for the door lights). In household units, it is assumed that there does not exist capacitive loads. The main goal behind a DSM program is to shift the on state of certain loads to off-peak hours and also to adjust the power drawn during this on state. To achieve this goal, the electric loads can be classified into three main classes, as follows: 1) Inelastic loads: Such loads are essential to meet the customers' comfort, and hence, shifting or reducing energy consumption of these loads are usually avoided. This class of loads includes lighting, TV, computers, refrigerators, and cooking appliances. 2) Shiftable loads: These loads are flexible in a sense that they can be deferred and rescheduled for operation during off-peak hours. The deferral period of the operation of such appliances depends on customer preference, and in general, elastic loads can be further categorized into two subgroups. The first subgroup of loads represents delay-sensitive appliances whose operation must be completed by a hard deadline, as in EV charging that should be complete within a specific duration set by the customer. The second subgroup represents delay-tolerant appliances that do not present hard deadlines that must be met, as in deferring the washing and drying of the customers' clothes until all other essential jobs are completed. 3) Adjustable loads: These loads have controllable power consumption pattern, as in air conditioners, where the amount of consumed power can be adjusted by controlling the target cooling temperature. Overall, shiftable and adjustable loads offer a trade-off between the customers' comfort level and the resulting reduction in consumed power.

DSM strategies can be broadly classified into two groups: price-based mechanisms and DLC mechanisms. In price-based mechanisms, different tariffs are adopted by the grid operator so that the customers are encouraged to shift and adjust their power consumptions from high peak period to off-peak periods. In this context, different pricing schemes are proposed in the literature, which include static pricing schemes as in time-of-use (TOU) pricing and critical peak pricing (CPP), and dynamic pricing schemes such as real-time pricing (RTP). Specifically, TOU rates divide the day into peak, mid-peak and off-peak periods and a fixed rate is calculated for each period. In CPP rates, the customers pay the highest price during critical load periods, and for the rest of the periods CPP acts exactly the same as TOU. In RTP, the tariffs are computed on an hourly basis and a pricing signal is delivered to the customers either an hour ahead or a day ahead. It is clear that price-based DSM mechanisms are based on the interaction between the customers and the utility where the customers respond to the expected high tariffs by reducing their peak load via shifting and/or adjusting their loads. On the other hand, DLC mechanisms enable the utility to directly control the customers' appliances through shifting and adjusting the energy consumption in return for monetary incentives that are offered to the customers. Hence, the customer signs an agreement (contract) with the utility company to enable them to control his/her appliances in exchange for a differentiated tariff for instance. Such a control should also respect the customers' preferences and comfort levels. Consider for instance periodic DLC events that control the operation of air conditioning appliances in a household during the summer months. From Figure 4.4a, the switch-off period of the air-conditioning appliances is 15 minutes, which is repeated periodically. As shown

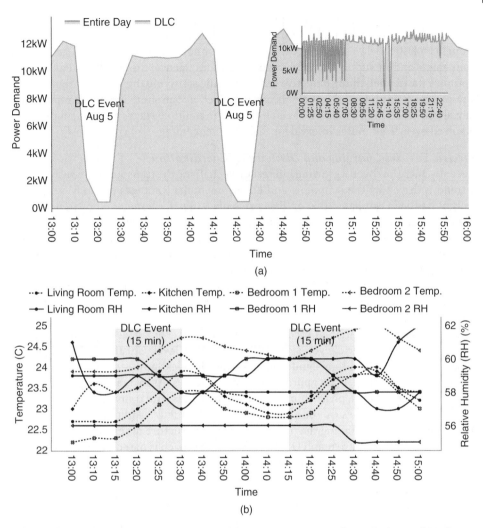

Figure 4.4 Periodic 15-minutes DLC events [9]. (a) Energy consumption during DLC event (b) Indoor temperature and humidity

in Figure 4.4b, the increase in the temperature as a result of the DLC switch-off events never exceeds 25 degrees due to the short DLC duration. The amount of saved energy can be high if such an event is repeated periodically over the day. Such an approach achieves a balance between the amount of saved energy and the customer's comfort level, which is measured by the corresponding increase in the temperature and humidity as a result of switching off the air conditioning appliances.

To design effective DSM strategies, an optimization problem is formulated. The objective function of this optimization problem should account for several factors such as: the electricity cost, distribution system losses, customer comfort level, carbon emission level, and reaching a target load profile. Hence, usually such problems are formulated as multi-objective optimization problems. Several constraints must be satisfied while solving the DSM optimization problem including: 1) the load constraints in terms of the lower and upper levels of the appliance power consumption, 2) the comfort level of the customer corresponding to the load deferral and/or adjustment, 3) load types, namely inelastic, delay-sensitive, delay-tolerant, or adjustable loads, and 4) generation constraints and/or supply-demand match constraint. Hence, the DSM strategy aims to optimize the defined multi-objective function while satisfying the aforementioned constraints. Such optimization problems can be solved in one shot, and hence represent a day-ahead scheduling problem, or can be solved in multiple stages where a subproblem is solved every 15 minutes interval for real-time scheduling.

Electric Vehicles Charging and Discharging Coordination

Recently, there has been a growing interest in electrifying the transportation sector. Such a trend is driven by the economic and environmental advantages of the EVs. Specifically, it has been estimated that an EV costs only $30 per kilometer compared with its gasoline-fueled counterpart that costs $173 per kilometer [10]. Furthermore, it has been reported that EVs that completely rely on rechargeable batteries can reduce the GHG emissions by 70%, which presents a positive environmental impact. The aforementioned advantages have motivated the wide deployment of EVs. It has been estimated that at least 500,000 high-capable EVs will be on Canadian roads by 2018 [11]. Furthermore, reports have indicated that the EV penetration level will reach 35%–62% in the period 2020–2050 [12].

In general, an EV charging process that is not effectively managed presents a potential risk to the power grid, even at low EV penetration level. Clustered EVs in specific geographical areas lead to a significant stress on the local power distribution system. Such an additional load, if not appropriately managed, results in increased power losses, power quality problems, phase imbalances, fuse blowouts, and transformer degradation [13, 14]. Two approaches can be adopted to deal with the EV additional load. The first approach relies on upgrading the power system infrastructure or installing DG units to satisfy the excess power demand [15]. The second approach is based on coordinated EV charging that controls the charging demand either by deferring the EV charging request or controlling the EV arrival rate to the charging facility [16]. The objective is to avoid the peak load coincidence for both regular loads (i.e., commercial, industrial, and residential loads) and EV loads. Coordinated EV charging is more appealing to the grid operators since it does not require big investment. In literature, two categories of charging coordination can be distinguished. The first category involves charging coordination among EVs at smart parking lots (in commercial, industrial, or residential areas). Hence, coordination takes place among a set of stationary (parked) EVs. In this context, coordination is adopted on a temporal basis. The second category involves charging coordination among mobile EVs that may require fast charging when moving on the road, such as electric buses and taxis. Hence, coordination is implemented on both spatial and temporal basis.

The main challenge when dealing with charging coordination for EVs within parking lots is how to anticipate the future EV charging demand for better temporal coordination. Specifically, in literature, the EV charging coordination decisions take into consideration the current and future regular loads and only the current EV charging demands and how long these EVs are expected to stay in the parking lot. Based on this information, a central vehicle controller (CVC) unit decides which EV is to be charged and which is to be deferred to a future time slot. However, better coordination can be achieved if the CVC unit has some insights regarding also the future EV charging demands. This approach can be realized via a smart real-time coordination system (SRTCS) [15]. The SRTCS enables both charging and discharging of EVs to allow energy exchange among vehicles, and hence better satisfies the customers' requests. It consists of three units. The first is a data collection unit that gathers information about the EV current power demands, the current EV battery SOC, the EV arrival and departure rates over the day within the parking lots (based on historical data), and the current and future regular loads. The second unit is a prediction unit that forecasts (with a small prediction error probability) the number of EVs that will be simultaneously present in the parking lot during the next time interval, given the gathered information. The last unit is an optimization unit that determines the charging and discharging coordinated decisions based on the current and future (anticipated) power demands of both EVs and regular loads. The coordinated decisions aim to maximize the customers' satisfaction while minimizing the system operating costs in terms of cost of power losses and peak demand charges (calculated based on the peak load reached within each time interval). Simulation results shown in Figure 4.5 for high EV penetration level demonstrate that the SRTCS reduces the EV load during the peak periods of the regular load (i.e., at 3:30 and 10:30pm) either using a charging only strategy (SRTCS-C) or by enabling a charging and discharging strategy (SRTCS-C/D). The EV load peak using the SRTCS occurs during the off-peak periods of the regular load. Define the strategy success factor as the average success of EV charging for all EVs in the system over the day. The SRTCS-C and

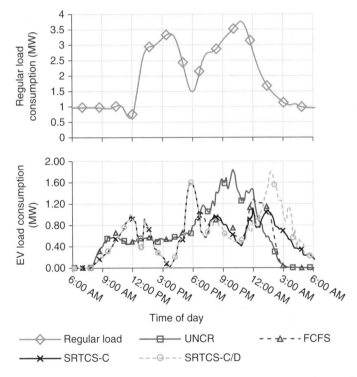

Figure 4.5 Performance of SRTCS charging and discharging coordination for parked EVs in comparison with uncoordinated charging and first come first served charging coordination [15].

SRTCS-C/D strategies have success factors of 91.3% and 97.7%, respectively. On the other hand, a first-come-first-served (FCFS) strategy has a success factor of 85% and the uncoordinated charging results in infeasible decisions that violate the power grid technical limitations (i.e., cause overload problems).

Mobile EVs, such as electric buses and taxis, may require fast charging while moving on the road. In this context, given a set of EVs and charging stations within a geographical region, it is required to assign EVs to the stations to get charged while avoiding overload problems. Consequently, another dimension is added to the coordination problem, namely, the spatial dimension in addition to the temporal one. Specifically, besides determining the appropriate charging instant for different EVs, it is required to determine the appropriate charging station. This is mainly due to the fact that different charging stations located at different buses experience both temporal and spatial fluctuations in their load capacity [17]. To enable the spatial and temporal charging coordination strategy, information regarding the fluctuating load capacities at the charging stations should be provided. Furthermore, the designed strategy should account for the EV real-time information such as its location and current SOC. This information is necessary to address the range anxiety problem that represents the tension between the EV travel cost (i.e., consumed energy on the road to reach the charging station) and the EV battery level. Overlooking the range anxiety problem may result in the EV battery being depleted on the way to the assigned charging station. A larger total-EV-charging energy (TECE) can be achieved, as shown in Figure 4.6, compared with a strategy that adopts only temporal coordination. The TECE gain can reach up to 35%.

The EV charging demand profile is mainly shaped by the customers' preferences, which can be controlled through economic incentives, as in DSM strategies. Consequently, price control can be adopted as an effective tool to align the EV charging

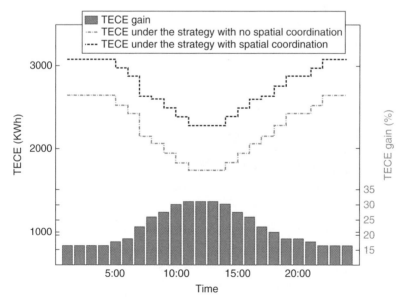

Figure 4.6 Performance of spatial and temporal charging coordination of mobile EVs in comparison with temporal only coordination [17].

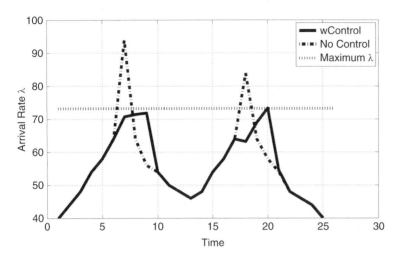

Figure 4.7 Demand shaping of EVs using price control strategy [18].

demands to the power grid available resources [18]. The charging station operator can calculate an optimal arrival rate that results in an acceptable blocking probability for the EV customers' charging requests. The charging station then regulates the charging requests arrivals to comply with the optimal arrival rate value by offering a discount to EV customers in return of postponing their charging requests to the next time interval. The EV customers that refuse to defer their charging requests may be subjected to service blocking. The optimal discount value is calculated such that it maximizes the charging station profit and satisfies the optimal arrival rate. The EV charging demand profile shaping through this strategy is shown in Figure 4.7 where the EV's two peak charging demands are offloaded to the next time intervals through incentives. This relieves the stress on the power grid and reduces the blocking probability of the EV charging requests.

Vehicle-to-vehicle (V2V) energy exchange through swapping stations offers an efficient approach to offload the EV high power demands and ensure overload avoidance. In this context, the main challenge for the V2V coordination strategy is twofold: 1) charging coordination for a set of EVs and 2) offering incentives to another set of EVs for discharging. The main objective is to ensure that the demand and supply are matched without requiring any additional (complement) energy from the power grid. Again, price control mechanisms can motivate EVs to contribute to a V2V energy transaction thanks to the expected low cost for charging EVs and high revenue for discharging EVs [19, 20].

4.2.2 Communication Requirements for Smart Grid Services

Table 4.1 presents a summary of the communication requirements of selected smart grid services for the collection of monitoring data, and transmission of pricing data and control signals. The communication requirements are expressed in terms of data size, latency, and reliability. Furthermore, the frequency of transmission is also highlighted. The data in Table 4.1 is adopted from [21]. As shown in Table 4.1, different services present high reliability and low latency requirements. This is especially true for

Table 4.1 A summary of communication requirements for different smart grid applications

		Application	Data Size	Sampling	Latency	Reliability
Monitoring Data Collection	Customer's Consumption	Communication between home appliances and customer's SM/energy management system	10-100 bytes	Once every configurable period (e.g., 15 minutes)	Seconds	> 98%
		On-demand customer's SM reading: from SM to utility	100 bytes	As needed	< 15 secs	> 98%
		Scheduled customer's SM reading: from SMs to AMI head end	1600-2400 bytes	4-6 times/day	< 4 hrs	> 98%
		Bulk customer's SM reading: from AMI head end to utility	MB	Multiple times/day for group meters	< 1 hr	> 99.5%
		EV charging status: from EV to utility	100 bytes	2-4 times/day	< 15 secs	> 98%
	Distribution & Txion	Distribution system monitoring: from line meters and field devices to utility	100-1000 bytes	1 time per device per hour	< 5 secs	> 99.5%
		Voltage stability monitoring	> 52 bytes	Once per 0,5 - 5 secs	< 30 secs	> 99.9%
		PMU-based state estimation: from PMUs to utility	> 52 bytes	Once every 10 milli-secs	< 10 milli-secs	> 99.9%
		Dynamic state estimation: from line meters to utility	> 52 bytes	Once every 0.2 -10 milli-secs	< 10 milli-secs	> 99.9%
Pricing Data Transmission		TOU: from utility to customer's SMs	100 bytes	1 time/device per price data txion event 4 times/year	< 1 min	> 98%
		CPP: from utility to customer's SMs	100 bytes	1 time/device per price data txion event 2 times/year	< 1 min	> 98%
		RTP: from utility to customer's SMs	100 bytes	1 time/device per price data txion event 6 times/day	< 1 min	> 98%
		Pricing to EVs: from utility to EVs	255 bytes	1 time per EV per 2-4 day	< 15 secs	> 98%
Control Signaling		DLC: from utility to customers	100 bytes	1 time per device per event	< 1 min	> 99.5%
		ESS Charge/discharge command	25 bytes	2-6 times/ dispatch period /day	< 5 secs	> 99.5%
		Distribution system demand response	150-250	1 time/device per 12 hrs	< 4 secs	> 99.5%

distribution and transmission system monitoring applications such as state estimation and voltage stability monitoring. In addition, transmission of control signals poses high reliability requirements. It should be noted that the amount of data to be transmitted presents a large volume due to the mass number of involved devices such as SMs, PMUs, EVs, etc.

As discussed in the introduction section, such high reliability and tight latency requirements cannot be satisfied by 4G networks. For instance, the state estimation application requirement of 10 milliseconds latency with 99.9% reliability requirements cannot be achieved when 4G networks are employed. Another major concern is how to guarantee that the requirements of these applications are simultaneously satisfied. On the other hand, 5G networks can address these limitations through a set of unique features such as massive connections, ubiquitous connectivity, ultra-low latency, ultra-high reliability, and very high throughput. These features can be achieved in 5G networks thanks to a group of novel concepts, namely, network slicing, NFV, SDN, and cloud computing. In the following, we briefly introduce these concepts for the sake of completeness, while more detailed information can be found in the introductory chapter of the book.

5G Fundamental Concepts

Network Slicing The concept of network slicing enables the network operator to slice (split) one physical network into multiple virtual networks, where each virtual network is optimized for a specific service. A key concept in network slicing is physical resource isolation, where the physical resources allocated to one network slice cannot be shared by other slices. Hence, multi-tenants co-exist on the same physical network in logical isolation. Cloud computing, NFV, and SDN are key enabling technologies to implement vertical slicing of the core, transport, and radio access networks and horizontal slicing of the terminals [22].

Cloud Computing mobile cloud computing (MCC) was introduced to act as an infrastructure as a service (IaaS) for data processing and storage outside the mobile device. This approach can also be regarded as an offloading technique of computation-intensive tasks from a resource-constrained device to a resource-sufficient cloud environment [23]. In terms of network slices, this is viewed as a horizontal slicing approach that helps devices to go beyond their physical limitations via resource sharing among high-capability network nodes and less-capable devices [22]. Unfortunately, MCC services offered through public clouds suffered from long response times, and hence failed to fulfill the latency requirements due to the centralized cloud architecture model. In turn, this affected the users' quality-of-service (QoS). To address such a limitation, the concept of edge cloud was introduced through Fog and Mobile Edge Computing.

Network Function Virtualization The idea behind NFV is to decouple the network functions and services from the hardware devices. As a result, network functions and services such as packet data network, serving gateways, firewalls, and load balancing can be implemented in software. One advantage of NFV is that network functions and services can be uploaded into cloud platforms, which enables savings in both capital and operational expenditures. Furthermore, NFV allows a high flexibility both in data and control planes, where any required upgrades and resource scaling can be implemented in software. The concept of NFV is a key enabler to vertical slicing of the core network

[22]. Specifically, NFV enables the co-existence of multi-tenants on the same physical hardware and splits the load balancing function from the hardware.

Software Defined Network The goal of SDN is to transform the decentralized network architecture into a centralized one by separating the packet forwarding process (data plane) from the routing decision making process (control plane). Such a decoupling makes the network switches become simple forwarding devices while the routing control actions are taken by a centralized controller that instructs the physical devices. This enables the network control to become directly programmable. As a result, the underlying network infrastructure can be abstracted from the network services. The SDN architecture consists of the following layers [24]: 1) application layer, which presents a set of programs that communicate their desirable network requirements and behavior to the SDN controller via a northbound interface (NBI), 2) control layer, which assigns instructions to the hardware devices (routers and switches) via a southbound interface (SBI) in accordance to the application layer requirements, and 3) physical layer, which consists of a set of hardware devices that follow the control layer instructions to accomplish the data transfer task. The concept of SDN is a key enabler to vertical slicing of the transport network [22]. Specifically, transport network slicing consists of slicing of both the control and forwarding planes. The control plane provides the necessary resource allocation for the forwarding plane, while the forwarding plane provides the QoS guarantee given the requirements of different network slices.

The above discussion illustrates that both SDN and NFV rely on virtualization to abstract the network infrastructures in software, which then will be implemented across the hardware platforms. In summary, SDN aims to separate the network control functions from the forwarding functions, while NFV abstracts network forwarding and other network functions from the hardware devices on which they run. Both concepts are useful to implement virtual networks that can share the same substrate network and operate independently and in isolation [25]. This is also known as external network virtualization, which splits one or more networks into a set of virtual networks. As shown in Figure 4.8, each substrate node can support a set of virtual nodes that will operate without interfering with each other.

4.3 Smart Grid Services Supported by 5G Networks

This section demonstrates how the 5G new concepts such as SDN, network virtualization, and cloud computing offer enhanced services for grid monitoring, data processing, DSM, and EV charging and discharging coordination. A summary of the application of these concepts in supporting the smart grid services is illustrated in Figure 4.9.

4.3.1 Data Collection and Management Services

This subsection is divided into two parts. The first part illustrates how to efficiently transmit the readings of SMs, PMUs, and sensors using concepts of SDN and network virtualization. The second part explains how to use cloud computing to share these readings among different operators, and hence, achieves a more accurate grid monitoring service, while detecting any malicious readings.

Virtual Network 1

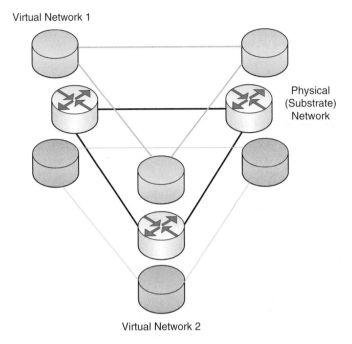

Physical
(Substrate)
Network

Virtual Network 2

Figure 4.8 Illustration of the network virtualization concept.

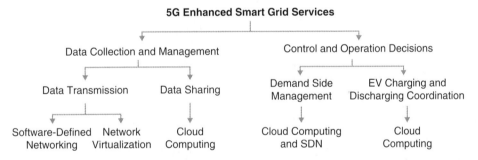

Figure 4.9 Summary of the application of 5G concepts in supporting smart grid services.

4.3.1.1 Data Collection Services
SDN-based Solutions As illustrated in Section 4.2, the future smart grid is characterized by massive number of SMs, PMUs, sensors, and actuators that are allocated at DG and ESS units, transmission and distribution lines, and inside the smart homes. In addition, some of the meters' sampling frequency is expected to be quite high (e.g., for PMUs as shown in Table 1). As a result, a huge volume of data will be generated at the future smart grid that needs to be efficiently routed in real time among various geo-separated entities. This poses some challenges related to managing the communication network resources. A traditional static communication network presents an inefficient solution to handle the real-time data acquisition process in the smart grid due to the expected high delay, poor bandwidth utilization, inefficient routing, and low throughput. To ensure a reliable data flow in real time over the underlying communication network, SDN presents an

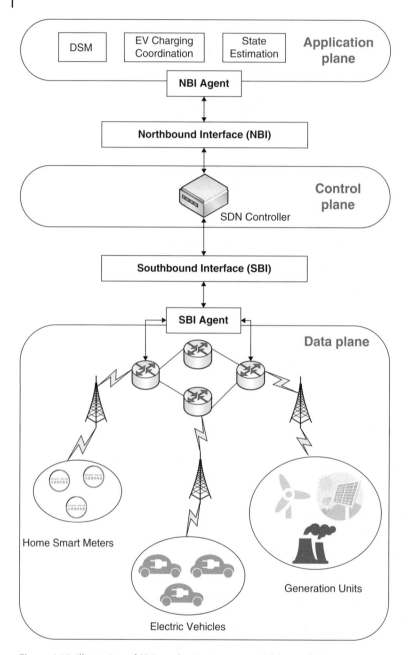

Figure 4.10 Illustration of SDN application in smart grid data collection.

effective solution due to its flexibility, programmability, and self-configuration. As the routing decision making is taken at a central controller in the SDN paradigm, fast data forwarding can be done in real time [26]. Figure 4.10 illustrates the application of SDN in supporting the smart grid data collection service. Thanks to the centralized control mechanism in SDN, optimal routing paths that ensure target QoS can be established.

Furthermore, efficient management of the network resources can be achieved by direct programmability and effective cross-layer design. Moreover, centralized control of the communication network via SDN offers enhanced capabilities against cyber-attacks by fast isolation of the compromised communication devices, efficient dropping of malicious traffic, and establishing on-demand paths in response to a cyber-attack [27].

In literature, two schemes are deployed to handle the smart grid big data in real time via SDN. The first approach integrates SDN with data dimensionality reduction techniques [28]. In the data plane, data is acquired from the SMs, PMUs, and sensors. The Frobenius norm is firstly applied to the collected data to generate reduced tensors with minimal reconstruction error for efficient data representation. By representing the smart grid big data using tensors, redundant and ambiguous dimensions can be removed, which enables a reduced transmission time. For instance, if a data frame with a size of 20 Mbits is transmitted at a data rate of 20 Mbps, it will reach the destination node in 1 second. Now, by reducing the data frame size to 16 Mbits and transmitting it with the same data rate, the transmission time will be reduced to 0.8 seconds. Upon decomposing the collected data into tensors of reduced size, the SDN controller (in the control plane) will dynamically determine the optimal routing paths that can achieve low latencies and high QoS. These routing decisions will be sent from the SDN control plane to the SDN data plane via the SBI, and thus the forwarding nodes at the data plane implement such routing decisions. A routing algorithm is presented in [28] for the SDN controller, using an empirical probability-based scheme that estimates the paths to forward the reduced data. The results in [28] show that the estimated path offers low latency and high throughput. Furthermore, the routing scheme in [28] is shown to achieve high route estimation accuracy even with increased packet losses. Moreover, the routing scheme in [28] achieves high bandwidth utilization and avoids any under-utilization of bandwidth or network congestion. Another routing algorithm is proposed in [29] based on an integer linear programming model, which is an NP-hard problem. Given a set of SMs, MDMSs, and switches, the objective is to route the generated data from the SMs to the appropriate MDMS via the switches. To address the associated computational complexity, the original problem is decomposed into two sub-problems, where the first sub-problem deals with the selection of the appropriate MDMS while the second sub-problem find the optimal path to reach the selected MDMS. The second approach that is adopted to handle the smart grid big data in real time based on SDN follows a hierarchical framework [24]. Generally, the distribution network of the power grid has two tiers, where the first tier carries the power from the substation to the primary feeders and the second tier carries the power from the primary feeders to the secondary feeders to supply the residential units. Hence, to achieve a good management of resources, the SDN-based hierarchical framework of [24] implements two-tiers as well. The SDN controller of the first tier is deployed at the substation, which aims to present a wide view of the system. This controller is responsible for routing the data from the SM aggregators to the substation at the primary network. The SDN controller of the second tier is implemented at the aggregator itself and aims to manage data collection from the SMs to reach the aggregator at the secondary network. Such a hierarchical implementation offers a flexible and efficient way to manage big data collection in the smart grid while offering global monitoring of the power grid.

It is shown in [29] that NFV can work with SDN to efficiently overcome any constraints in resources while processing the smart grid big data in the MDMSs. Specifically, the networking resources such as switches and routers will be managed by the SDN controller while all other resources required for processing and storing the data will be managed by the NFV orchestration. In this context, virtual MDMSs will be created and a set of hypervisors will manage the resource split among these virtual MDMS and the SDN controller will determine the forwarding tables for all switches such that the SM data can be efficiently routed to the appropriate virtual MDMS. Such an integration between SDN and NFV offers a customizable and programmable virtualized MDMS that can be efficiently scaled according to the incoming SM traffic load.

Network Virtualization-based Solutions Different types of communication networks can be employed for data transmission in the distribution grid such as wireless mesh network (WMN), power line communication (PLC), cellular networks, and vehicular ad-hoc networks (VANETs). However, these networks may suffer from packet loss and bit-error-rate due to interference and attenuation. Hence, a single network cannot guarantee real-time service support with low latency and high reliability as required for smart grid applications. Using network virtualization, it is possible to integrate different heterogeneous networks for efficient data collection with high QoS guarantee. For instance, in [25], smart grid real-time services are supported by virtual networks that are mapped to WMN and PLC. Through virtual network embedding, virtual nodes and links are mapped onto the substrate nodes and paths. To achieve the high-reliability requirement, real-time services were supported in [25] by virtual networks that are mapped to both WMN and PLC via concurrent transmission, multiple subcarrier allocation, and time diversity. On the other hand, a single network mapping is adopted in [20] to support data collection from mobile EVs regarding the battery SOC, power charging demands, and EV current location. In this framework, the EV first uses the VANET for data transmission when the connection probability is estimated to be larger than a predefined threshold, e.g., 80%. When the connection probability in the VANET is below this threshold, the EV employs the cellular network to transmit the real-time data. Such a variation in the VANET connection probability can be due to a change in the vehicle density or a change in the transmission range. Hence, the network integration solution proposed in [20] makes use of the VANET low transmission cost and relies on the cellular network only when poor connections are expected in the VANET, in order to guarantee successful information transmission. Furthermore, such an approach helps in relieving possible congestions induced by EV transmissions over the cellular network.

Cloud Computing-based Solutions Data transmission in smart grids are mostly carried out in a hierarchical architecture, where data is first transmitted from SMs to aggregators and then aggregators forward this data to the smart grid control center. One way to efficiently implement this aggregation is based on Fog cloud nodes. In this paradigm, the grid is divided into a set of regions/microgrids and a Fog node is assigned to each microgrid. Data is transmitted from the SMs in each microgrid to the Fog node in charge of this microgrid. Each Fog node will implement in-network aggregation and then forward this data to the smart grid control center, which could be implemented in the core cloud. However, a serious concern that is raised in this context is related to preserving the privacy of the customer's data. In order to minimize the privacy leakage, Gaussian noise addition is employed at the Fog nodes to present a differential privacy guarantee

[30]. This technique of noise addition has also been investigated in [31], and the results indicate that such an approach improves the data anonymity and presents a minimal impact on the optimality of the decisions to be taken using the noisy measurements. Furthermore, to ensure the aggregator obliviousness, data encryption is adopted in [30] and both encryption and permutation techniques are adopted in [31].

4.3.1.2 Data Management Services

Cloud Computing-based Solutions In a liberated electricity market, the power grid is divided into regions and each region is monitored and managed by an independent system operator (ISO) or a regional transmission organization (RTO) that coordinates the transmission utility companies to ensure reliable, efficient, and secure regional power system. Each ISO manages a set of PMUs and sensors that are used to monitor its region of the grid and implements a separate state estimation to observe the voltages at different nodes and ensure they do not violate the minimum and maximum thresholds. However, to enable coordination among such ISOs, wide area monitoring is required. As a result, data sharing among ISOs should be carried out in a secure, scalable, and cost-effective way with low latency for real-time service support. An open source cloud computing platform, namely GridCloud, is proposed in [32] to implement such a data sharing objective in the most efficient manner. Data from PMUs, SMs, and sensors of different ISOs and utility companies flow into a shared cloud-computing platform for storage and processing. State estimation task can be carried out using the shared data from all collaborating entities to present a wide visualization of the system state, and hence, better coordination can be achieved among different regions. ISOs and utility companies can draw the shared data and analytic outputs to better manage their assets. The GridCloud platform offers several advantages such as consistent data availability at a massive scale. High availability is achieved at GridCloud as data is saved redundantly over multiple servers. Furthermore, redundant connections are also established to ensure reliability in response to communication link failures.

In addition to supporting data sharing services, cloud computing presents an effective tool to mitigate cyber-security attacks. Since a massive number of SMs will be deployed in the customers' premises, these meters are considered to be unprotected devices that are distributed in the low-trust environment. Being the primary access to the AMI network, such SMs pose serious security threats. Specifically, any manipulation of the data transmitted through these SMs directly impacts many services in the smart grid such as consumption monitoring, DSM, and billing. In this context, cloud computing can provide cyber-security solutions via a security-as-a-service (SECaaS) model. The work in [33] proposes a cloud-centric collaborative monitoring architecture that involves two major participants. The first participant is the smart grid operator that owns an AMI network to collect data from the SMs for monitoring and DSM purposes and the second participant is a SECaaS provider that offers network security monitoring, encryption key management, and security assessment. At the smart grid operator side, monitoring and decision entities are established to perform initial screening of data for assessment against security attacks, such as data manipulation attacks, based on a set of predefined criteria. After this initial screening, data is transferred to a cloud computing platform that consists of a set of powerful servers that will run advanced analytical tools, e.g., deep machine learning techniques. For instance, a deep feed-forward neural network is designed in [7] to efficiently detect data manipulation attacks in AMI networks. Detection latency represents a very important performance evaluation metric. This latency is translated to a time difference between detecting a cyber-attack event and taking the appropriate response. In order to minimize the overall detection latency, a placement problem is formulated in [33] to find the optimal location for the cloud servers in order to minimize the access link latency, and hence, the overall latency.

4.3.2 Operation Decision-Making Services

This subsection is also divided into two parts. The first part explains how to employ the concepts of cloud computing and SDN in supporting decision making for DSM. The second part focuses on employing these concepts in supporting decision making for EV charging coordination.

4.3.2.1 Demand Side Management Services

Cloud Computing-based Solutions Usually, DSM programs run in a distributed fashion via iterative algorithms. Customers are assumed to belong to different clusters (microgrids). Each customer within a specific microgrid solves locally an optimization problem and then broadcasts some information update to other customers in the microgrid to update global parameters for this microgrid. This procedure is repeated until all customers converge to the global optimal decision, which can be in form of appliance scheduling decisions. There are several implications with this approach. First, decisions about dynamic pricing and coordination among different microgrids need to be taken in real-time. When the communication channel among the customers is not ideal, the update messages that are exchanged among the customers might be lost. Using transmission control protocol (TCP) in transferring the update messages will increase the convergence time to reach the global optimal solution, as the same update message will be re-transmitted several times until an acknowledgment message is received from all customers. On the other hand, by adopting a user datagram protocol (UDP), the delivery of the update messages is not guaranteed. As a result, the DSM optimization problem might converge to a non-optimal solution as outdated update messages will be used at some customers. Second, high computational power is required to solve the DSM optimization problem in a reduced time. The local problems to be solved at each customer will be handled by the customer's home energy management system (HEMS) or smart meter, which does not present the required computational, memory, and storage capabilities to tackle such optimization problems. Cloud computing-based solutions offer a semi-centralized mechanism that can effectively address the aforementioned challenges. These solutions assume a two-tier cloud computing architecture consisting of an edge cloud (or a Fog cloud) and a core cloud [34–36]. Customers within each microgrid are controlled by an edge (Fog) cloud to provide the required computational and storage resources that can meet the desirable low latency. The core cloud solves a centralized optimization problem to coordinate the operation of different microgrids. Energy consumption data, local generation from DG units, and available energy at ESS units are transmitted from the customers to the edge cloud, which solves an optimization problem to determine the optimal power consumption schedule for the customers' appliances and ESS units. Each edge cloud then forwards the total load at each microgrid to the core cloud, which will solve a

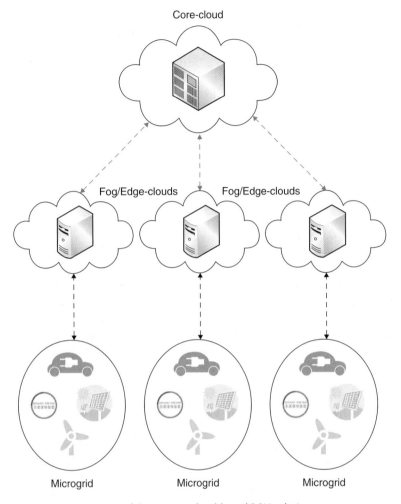

Figure 4.11 Illustration of the two-tier cloud-based DSM solution.

centralized problem to coordinate the resource usage among the microgrids in order to minimize the overall cost. Figure 4.11 illustrates the two-tier cloud-based solution for DSM. This two-tier cloud-based solution presents several advantages as follows: 1) The semi-centralized approach alleviates the challenges related to the wireless channel quality, in terms of packet loss. The customers only forward their data to the edge cloud and the edge cloud is responsible for solving the scheduling optimization problem, and hence, issues related to missing or outdated update messages due to a poor channel quality is no longer a concern. 2) The two-tier approach offers the high computational power that is required to solve the DSM optimization problems since all the calculations are done at the edge and core clouds without any computational burden on the HEMS or SM at the customer premises. 3) This approach also provides the necessary storage capabilities at the cloud servers and reduce the volume of traffic that is required to traverse the communication network backbone. 4) A major advantage provided by the cloud-based solutions is that they can leverage the cloud defense mechanisms against privacy and security cyber-attacks, unlike the traditional peer-to-peer distributed DSM solutions that are vulnerable to cyber-security attacks and also may violate the customers' privacy.

The concept of SDN can be well integrated with cloud computing to offer a flexible and scalable DSM service in smart grid [37]. In this context, all communication networks that are responsible for collecting the meter readings are managed and controlled via SDN to ensure real-time information delivery for decision making, while the cloud data servers handle the DSM decision making process. The SDN controller can be implemented inside the cloud to manage the backhaul network. The SDN controller supervises the communications between the microgrids and the edge (Fog) cloud. It is shown in [37] that when the power consumption requests increase in the grid, the Fog-SDN mechanism results in a reduced response time compared with a centralized core network infrastructure. This is due to the fast access provided by the distributed Fog nodes and the fast data forwarding actions carried out by the SDN.

4.3.2.2 Electric Vehicle Charging and Discharging Services

Cloud Computing-based Solutions Charging and discharging coordination can be implemented in a distributed manner where several EVs engage in an iterative algorithm to reach global scheduling decisions [20]. However, these EVs might not have the computational power, memory, and storage capabilities that are required to participate in solving such a large scale optimization problem. Solving these coordination problems in a centralized manner might not also be efficient, as high latency could be incurred when solving the coordination problem by a control unit that is located farway from moving EVs. One solution to address this problem is achieved by using mobile edge computing to offload the complex computations to be processed at the network edge, which also reduces the associated latency. A four-layer framework in proposed in [38] to coordinate charging and discharging requests of EVs in an efficient manner. The lowest layer includes the EVs and the second layer consists of routers and gateways deployed in the network backbone. The third layer includes a set of servers that are distributed at different geographical locations at the edge of the network, and the topmost layer represents the core-cloud infrastructure. The charging and discharging requests will be routed from the EVs at the first layer by the routers and gateways in the second layer to reach the nearest data server at the network edge to determine the optimal decision. Coordination among different data servers is implemented at the core-cloud to specify real-time pricing information from the owner of the charging station to the EVs.

The framework in [38] does not take into consideration service differentiation between moving and parked EVs. While moving EVs must be served in real time, parked EVs can tolerate longer delays. The framework presented in [39] accounts for this fact and presents a hierarchical cloud-based EV charging coordination model that involves two levels of cloud computing infrastructures that can meet different latency requirements of different customers. Specifically, two types of clouds are defined. The first type is referred to as local cloud, which presents shorter end-to-end delay, while the second type is a remote cloud that presents a relatively longer end-to-end delay. Two classes of services are considered, namely, high priority (HP) and low priority (LP) charging service. Mobile EVs are offered high priority service as they require real-time service, while parked EVs are offered low priority service as they have comparably longer latency requirement. The EVs in HP and LP services communicate with the local and remote clouds, respectively. The remote cloud aims to collect the information from the EVs in parking lots and synchronize the data with the local cloud to reach optimal charging coordination decisions.

4.4 Summary and Future Research

In this chapter, the application of 5G novel concepts in supporting advanced smart grid services is discussed. Concepts of SDN and NFV find great application in data collection from the massive number of SMs, PMUs, and sensors that are deployed across the smart grid. Through these two concepts, flexible and dynamic routing of big volume of data among various geo-separated locations can be achieved in real time. Furthermore, the concept of cloud-computing, especially Fog and mobile edge clouds, find great application in data management and decision making in smart grids. Specifically, efficient and secure sharing of sensor data is possible among various ISOs via the cloud, which enables enhanced monitoring services for the grid, and hence, more informative decisions can be taken. In addition, cloud-computing is very useful in offloading the complex computations that are needed in order to reach optimal scheduling and coordination decisions in real time for DSM and EV charging and discharging.

The research work in supporting smart grid services via 5G network concepts is still in the early stage and further investigations are required to several open challenges, as follows:

- The existing research works mainly focus on smart services that are still considered to be relatively delay-tolerant if compared with other emergency services. For instance, the self-healing capability of the smart grid requires fast detection, localization, and isolation of faults within the grid while ensuring continuity of power flow. This kind of emergency services poses more stringent latency requirements than DSM and EV charging and discharging coordination decisions. Further attention should be given to design effective communication platforms based on the novel 5G concepts to support the tight delay constraints of emergency services.
- The existing research focuses mainly on a single application scenario in smart grid and design the communication network based on the requirement of this application. However, the developed frameworks are not investigated when competition on resources exists among different applications. For instance, how can we dynamically schedule the network resources among different slices, each supporting a specific smart grid service, e.g., data collection, DSM operation, and EV charging and discharging coordination, while satisfying their QoS requirements? Static allocation of resources among these services might result in over- or under-provision situations, which either lead to inefficient utilization of the available resources in case of over-provision or negatively affect the smart grid services in case of under-provision. More importantly, how can we schedule the network resources among slices of different industries, e.g., smart grid, health care services, etc? How can we guarantee that these services meet their QoS requirements and ensure proper isolation among them, especially in a peak load network condition? Hence, the investigation of the developed communication platforms for smart grid services need to be investigated in a more realistic network setting that involves supporting different industries at peak network conditions.

- Further investigations are required on the application of machine learning-based techniques for network management. For instance, dynamic routing in SDN controllers can be designed using advanced machine learning tools, which enables more intelligent routing mechanisms that can easily and quickly adapt to the network conditions and the smart grid service requirement. Moreover, the management of virtual network resources can be carried out more efficiently using machine learning-based tools.
- Future research should focus on finding robust solutions to the challenging security threats that face the future 5G networks, which hinder their application in supporting the communication requirements of critical infrastructures such as the smart grid. Attackers may exploit the capacity elasticity of one network slice to consume the network resources that may be used to support smart grid services. Furthermore, effective security measures must be ensured while performing power grid applications on the cloud. Moreover, the existence of a single SDN controller may represent a security challenge for the data routing services within the smart grid, as it represents a single point of failure that may jeopardize the data collection service within the smart grid if the SDN controller was subject to a security attack.

Acknowledgment

This publication was made possible by NPRP grant # NPRP10-1223-160045 from the Qatar National Research Fund (a member of Qatar Foundation). The statements made herein are solely the responsibility of the authors.

References

1 M. Cosovic, A. Tsitsimelis, D. Vukobratovic, J. Matamoros, and C. Antón-Haro. 5G mobile cellular networks: enabling distributed state estimation for smart grids. *IEEE Communications Magazine*, 55:62–69, 2017.

2 China Telecom, State Grid, and Huwai. 5G network slicing enabling the smart grid. *Technical Report*, page 28, 2018.

3 N. Kayastha, D. Niyato, E. Hossain, and Z. Han. Smart grid sensor data collection, communication, and networking: a tutorial. *Wireless Communications and Mobile Computing*, 14:1055–1087, 2014.

4 X. Fang, S. Misra, G. Xue, and D. Yang. Smart grid - the new and improved power grid: a survey. *IEEE Communications Surveys and Tutorials*, 14: 944–980, 2012.

5 O. Alrawi, I. S. Bayram, and M. Koc. High-resolution electricity load profiles of selected houses in Qatar. *IEEE 12th International Conference on Compatibility, Power Electronics and Power Engineering (CPE-POWERENG 2018)*, page 6, 2018.

6 Q. Sun, H. Li, Z. Ma, C. Wang, J. Campillo, Q. Zhang, F. Wallin, and J. Guo. A comprehensive review of smart energy meters in intelligent energy networks. *IEEE Internet of Things Journal*, 3:464–479, 2016.

7 M. Ismail, M. Shahin, M. F. Shaaban, E. Serpedin, and K. Qaraqe. Efficient detection of electricity theft cyber attacks in AMI networks. *IEEE Wireless Communications and Networking Conference (WCNC)*, page 6, 2018.

8 I. S. Bayram and M. Koc. Demand side management for peak reduction and PV integration in Qatar. *IEEE 14th International Conference on Networking, Sensing and Control (ICNSC)*, page 6, 2017.

9 I. S. Bayram, O. Alrawi, H. Al-Naimi, and M. Koc. Direct load control of air conditioners in Qatar: an empirical study. *6th International Conference on Renewable Energy Research and Applications*, pages 1007–1012, 2017.

10 TESLA Motors. http://www.teslamotors.com/models.

11 Natural Resource Canada. Electric vehicle technology roadmap for canada. page 4, 2010.

12 Electric Power Resech Institute. http://www.epri.com/Pages/Default.aspx.

13 R. Liu, L. Dow, and E. Liu. A survey of PEV impacts on electric utilities. *Innovative Smart Grid Technologies (ISGT)*, pages 1–8, 2011.

14 M. F. Shaaban, M. Ismail, and E. El-Saadany. PEVs real-time coordination in smart grid based on moving time window. *IEEE PES General Meeting*, pages 1–5, 2014.

15 M. F. Shaaban, M. Ismail, E. El-Saadany, and W. Zhuang. Real-time PEV charging/discharging coordination in smart distribution systems. *IEEE Transactions on Smart Grid*, 5:1797–1807, 2014.

16 H. Liang, A. K. Tamang, W. Zhuang, and X. Shen. Stochastic information management in smart grid. *IEEE Communications Surveys and Tutorials*, 16:1746–1770, 2014.

17 M. Wang, M. Ismail, X. Shen, E. Serpedin, and K. Qaraqe. Spatial and temporal online charging/discharging coordination for mobile PEVs. *IEEE Wireless Communications*, 22:112–121, 2015.

18 I. S. Bayram, M. Ismail, M. Abdallah, K. Qaraqe, and E. Serpedin. A pricing-based load shifting framework for EV fast charging stations. *IEEE International Conference on Smart Grid Communications (SmartGridComm)*, page 6, 2014.

19 M. Wang, R. Ismail, Zhang, M., X. Shen, E. Serpedin, and K. Qaraqe. A semi-distributed V2V fast charging strategy based on price control. *IEEE Global Communications Conference*, page 5, 2014.

20 M. Wang, M. Ismail, R. Zhang, X. Shen, E. Serpedin, and K. Qaraqe. Spatio-temporal coordinated V2V energy swapping strategy for mobile PEVs. *IEEE Transactions on Smart Grid*, 9:1566–1579, 2018.

21 M. Kuzlu, M. Pipattanasomporn, and S. Rahman. Communication network requirements for major smart grid applications in HAN, NAN and WAN. *Computer Networks*, 67:74–88, 2014.

22 End to end network slicing. *White Paper*, page 40, 2017.

23 T. Taleb, K. Samdanis, B. Mada, H. Flinck, S. Dutta, and D. Sabella. On multi-access edge computing: a survey of the emerging 5G network edge cloud architecture and orchestration. *IEEE Communications Surveys and Tutorials*, 19:1657–1681, 2017.

24 N. Chen, M. Wang, N. Zhang, X. Shen, and D. Zhao. SDN-based framework for the PEV integrated smart grid. *IEEE Network*, 31:14–21, 2017.

25 P. Lv, X. Wang, Y. Yang, and M. Xu. Network virtualization for smart grid communications. *IEEE Network*, 31:14–21, 2014.

26 R. Chaudhary, G. S. Aujla, S. Garg, N. Kumar, and J. Rodrigues. SDN-enabled multi-attribute-based secure communication for smart grid in IIoT environment. *IEEE Transactions on Industrial Informatics*, 14: 2629–2640, 2018.

27 D. Jin, Z. Li, C. Hannon, C. Chen, J. Wang, M. Shahidehpour, and C. W. Lee. Toward a cyber resilient and secure microgrid using software-defined networking. *IEEE Transactions on Smart Grid*, 8:2494–2504, 2017.

28 D. Kaur, G. S. Aujla, K. Kumar, A. Zomaya, C. Perera, and R. Ranjan. Tensor-based big data management scheme for dimensionality reduction problem in smart grid systems: SDN perspective. *IEEE Transactions on Knowledge and Data Engineering*, page 14, 2018.

29 A. Montazerolghaem, M. Moghaddam, and A. Leon-Garcia. OpenAMI: software-defined AMI load balancing. *IEEE Internet of Things Journal*, 5: 206–218, 2018.

30 L. Lyu, K. Nandakumar, B. Rubinstein, J. Jin, J. Bedo, and M. Palaniswami. Ppfa: privacy preserving Fog-enabled aggregation in smart grid. *IEEE Transactions on Industrial Informatics*, page 11, 2018.

31 A. Sherif, M. Ismail, M. Mahmoud, K. Akkaya, E. Serpedin, and K. Qaraqe. Privacy preserving power charging coordination scheme in the smart grid. *Transportation and Power Grid in Smart Cities:Communication Networks and Services,Wiley*, page 26, 2018.

32 D. Anderson, T. Gkountouvas, M. Meng, K. Birman, A. Bose, C. Hauser, E. Litvinov, X. Luo, and F. Zhang. GridCloud: infrastructure for cloud-based wide area monitoring of bulk electric power grids. *IEEE Transactions on Smart Grid*, page 10, 2018.

33 M. Hasan and H. Mouftah. Cloud-centric collaborative security service placement for advanced metering infrastructures. *IEEE Transactions on Smart Grid*, page 9, 2018.

34 M. Yaghmaee and A. Leon-Garcia. A Fog-based Internet of energy architecture for transactive energy management systems. *IEEE Internet of Things Journal*, page 15, 2018.

35 M. Yaghmaee, M. Moghaddassian, and A. Leon-Garcia. Autonomous two-tier cloud-based demand side management approach with microgrid. *IEEE Transactions on Indutrial Informatics*, 13:1109–1120, 2017.

36 M. Yaghmaee, A. Leon-Garcia, and M. Moghaddassian. On the performance of distributed and cloud-based demand response in smart grid. *IEEE Transactions on Smart Grid*, page 15, 2018.

37 D. A. Chekired, L. Khoukhi, and H. Mouftah. Decentralized cloud-SDN architecture in smart grid: a dynamic pricing model. *IEEE Transactions on Industrial Informatics*, 14:1220–1231, 2018.

38 N. Kumar, S. Zeadally, and J. Rodrigues. Vehicular delay-tolerant networks for smart grid data management using mobile edge computing. *IEEE Communications Magazine*, 54:60–66, 2016.

39 C. Kong, B. Rimal, B. Bhattarai, I. S. Bayram, and M. Devetsikiotis. Cloud-based charging management of smart electric vehicles in a network of charging stations. *IEEE Transaction on Smart Grid, Under Review*, page 8, 2018.

5

Evolution of Vehicular Communications within the Context of 5G Systems

Kostas Katsaros and Mehrdad Dianati

5.1 Introduction

Connected and autonomous vehicles (CAVs) incorporate a range of different technologies with the aim of moving people and goods on our roads more safely and efficiently. CAVs, coupled with intelligent road infrastructure, will tackle some serious environmental, social and economic issues which will only get worse if no action is taken. Currently, the number of vehicles on our roads continues to rise. As of June 2017 there were 37.8[1] million licensed vehicles in the UK, up 41.7% since 1997. Of the total, 31.2 million were cars, up 38.6% over the 20 year timeframe. Congestion and pollution are already serious problems across most major towns and cities.

Over the last few years, the evolution of Advanced Driver Assistance Systems (ADAS) is progressively shifting towards connected and automated driving. With the promise of improving road safety and traffic efficiency in future Intelligent Transportation Systems (ITS), evolved driving features like Cooperative Adaptive Cruise Control (CACC) and Automated Platooning will require real-time vehicle information exchange with the surrounding environment, encouraging the development of highly reliable and low-latency vehicular communication systems. To that end, the National Highway Traffic Safety Administration (NHTSA) in the U.S. announced in 2016 a mandate to use Dedicated Short Range Communication (DSRC) for vehicle-to-vehicle (V2V) communications. This was subsequently retracted after a significant response from a wide range of stakeholders on alternatives, specifically cellular-based vehicle communications (C-V2X). In the European Union (EU), the Cooperative ITS (C-ITS) initiative still is undecided on which communication technology to select. However, the upper protocol stack and service deployment have a clearer roadmap. Recently, news reported[2] that the draft legislation is to promote DSRC over C-V2X, but that could potentially be disputed like the NHTSA mandate. The 5G Automotive Association published an assessment of the two primary technologies in 2017, recommending that the EU should not take a preference on any of the two solutions so as not to hinder the deployment of one in favour of the other for the provision of direct communications among vehicles and between vehicles and vulnerable road users. On the other hand, China has a clear direction towards

1 www.gov.uk/government/statistics/vehicle-licensing-statistics-april-to-june-2017
2 Volkswagen a winner as EU set to favour WiFi over 5G: draft https://reut.rs/2OuPkLQ

Enabling 5G Communication Systems to Support Vertical Industries, First Edition.
Edited by Muhammad Ali Imran, Yusuf Abdulrahman Sambo and Qammer H. Abbasi.
© 2019 John Wiley & Sons Ltd. Published 2019 by John Wiley & Sons Ltd.

C-V2X, with a target of 90% coverage by 2020. Similarly, the automotive industry is divided. Companies like VW are tending towards DSRC, Ford indicates that it will get cellular-based solutions, while BMW is taking a more radical approach with a hybrid DSRC/C-V2X solution. By 2025, CAVs will upload over 1 terabyte of vehicle and sensor data per month to the cloud according to Gartner. This is up from from 30 gigabytes from advanced connected cars in 2018. Further, with the increase of autonomy, drivers will turn to passengers that would consume even more data through high-quality infotainment services. Such a demand can only be supported by future 5G communications.

In this chapter, we investigate the evolution of vehicular communication systems towards 5G and how the applications and services follow that evolution. Specifically, in section 5.2 of this chapter, we are primarily investigating the cellular-based solution, and how it is evolving from LTE Release 14, the initial C-V2X system, towards Release 16, the fully-fledged 5G system. We are also touching upon the co-existence issues with DSRC and what other technologies contribute to efficient V2X services. Then, in Section 5.3 we are focusing on the data dissemination on top of a vehicular communication platform that could support efficient cloud-based ITS services. Further, in Section 5.4, we examine how the evolution of V2X communication technologies is mirrored on the evolution of services it supports, from awareness to autonomous driving. Finally, Section 5.5 concludes this chapter.

5.2   Vehicular Connectivity

Vehicular communications can be divided into (a) on-board, (b) off-board local (vehicle-to-vehicle and vehicle-to-pedestrian (V2P)) and (c) off-board remote (vehicle-to-network (V2N)) connections. There are all referred as V2X, vehicle-to-everything. In this section, we are only investigating off-board communications. In recent years, two candidates for off-board vehicular communications have evolved for the support of road safety and traffic efficiency applications. On the one hand, ad-hoc networks exist based on the DSRC, and on the other hand, there are cellular network infrastructures based on an extended LTE stack, which is commonly known as C-V2X or LTE-V. A recent survey on the regulation and open research challenges for V2X is available in [1].

Both approaches meet the requirements on vehicular communications but show technology-inherent mechanisms that result in different performances. DSRC features a small latency at a small network load whereas C-V2X promises a highly reliable packet transmission. One of the main difference of both approaches lies in the channel access which is random-based for DSRC and centrally scheduled for C-V2X. A disadvantage of LTE for V2X is the non-exclusive usage of the spectrum, when not operating in the 5.9GHz ITS-specific band. CAVs will have to share it with other mobile users, that can potentially cause performance degradation in highly populated areas. However, the wide adoption of cellular technologies both in automotive industry as well as other road users (e.g. pedestrians, cyclists), makes it one of the prime candidates for connected vehicles. Other benefits of C-V2X over DSRC include (a) a better coverage for infotainment and Internet access by exploiting existing cellular network coverage, especially when using lower frequency spectrum, (b) reduced infrastructure deployment cost and improved reliability by leveraging existing infrastructure and economics of scale,

(c) enhanced security through the user of mobile SIM cards. Further, the continuous evolution of LTE systems towards the next generation cellular communications (5G), will enable more sophisticated services and assist V2X applications such as autonomous operations with tighter latency requirements. Several research works can be found in the literature that analyse and compare the performance of the two technologies such as [2–4]. In the following sections we provide a qualitative review of the two technologies and their evolution towards future vehicular connectivity services.

5.2.1 Cellular V2X

The cellular approach is promising because the managed and therefore collision-free channel access, based on Frequency Division Multiplexing (FDM), results in a more reliable packet transmission, which is crucial for safety applications. Much effort has been invested into adopting cellular technologies for V2X communication purposes. The most important extension of the classical LTE stack is a direct link among user devices, which is also called Device-to-Device (D2D) communication. Since 3GPP Release 12 (Rel.12) and the introduction of D2D, many features have been added to the standard aiming to realize V2X services. The new interface for D2D communications is referred as PC5. The Proximity Service function introduced in the architecture has the following main roles: (a) resource provisioning, (b) direct discovery management and (c) Evolved Packet Core (EPC) discovery.

In 2014, 3GPP conducted studies within Rel.13 to test the applicability of LTE to V2X use cases. This has resulted in the specification document TS 22.185 on V2X use cases and requirements, published in March 2016. It was followed by the proposed architectural enhancements to support such services in Release 14 [5]. The work from 3GPP on C-V2X has been split in three phases focusing on different use cases, namely Phase 1 in Rel.14 for Basic Safety, Phase 2 with Rel.15 for Enhanced Safety and Phase 3 with Rel.16 looking towards autonomous driving.

5.2.1.1 Release 14 – *First C-V2X Services*
Phase one of 3GPP activities on C-V2X is addressing what is commonly known as Day-1 C-ITS application, which includes forward collision warning, emergency stop warning, pre-crash warning, traffic management and other issues. The V2X features encompass all aspects of the 3GPP work needed to support vehicle-based communications: enhancements of the air interface, protocols, and impacts on the LTE core network. There are two modes of operation for V2X communication, as shown in Figure 5.1:

- *V2X Communication Over PC5 Interface*: PC5 interface directly connects UEs (User Equipments) so that an over-the-air V2X message from a UE is directly received by UEs around the transmitter. It is possible for the network to schedule the resources (Mode 3) or the UEs to autonomously self-manage them (Mode 4). For *Mode 3*, the UE has to be in RRC_CONNECTED state to be able to transmit and requests resources from the eNB. These are scheduled for transmission of Sidelink Control Information (SCI) and data. In order to assist the eNB to provide sidelink resources, a UE in RRC_CONNECTED may report geographical location information to the eNB. For *Mode 4*, the UE selects resources from resource pools and performs

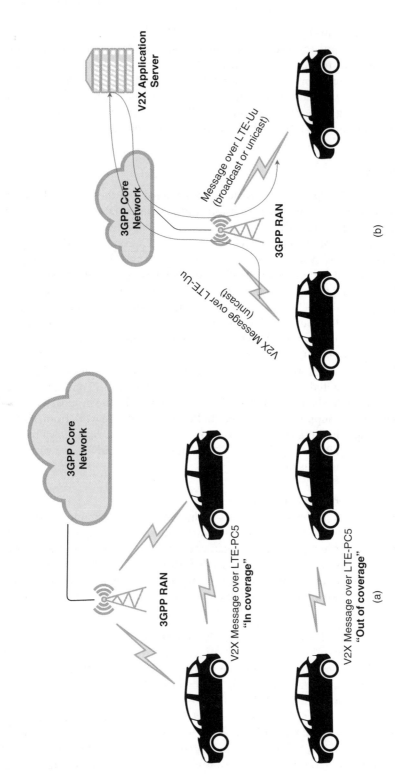

Figure 5.1 V2X communications in 3GPP Rel.14 (a) V2X communication over PC5 interface (b) V2X communication over LTE-Uu interface.

transport format selection to transmit SCI and data autonomously. Autonomous resource selection sensing before transmission with semi-persistent scheduling (SPS) is usually employed, which applies real-time measurements of the received power, with the aim of keeping a tolerable level of interference, while increasing the reuse of resources. Often, the resource mapping is pre-defined on specific geographical zones that can be configured by the eNB or preconfigured.

- *V2X Communication Over LTE-Uu Interface*: LTE-Uu interface connects UEs with the eNB which plays the role of base station in the LTE networks. This mode is mostly used for latency-tolerant traffic such as informational safety, telematics and infotainment services.

For the PC5-based communications, a new V2X control function is added and used for network related actions required for V2X, such as authorization from a public land mobile network (PLMN), as illustrated in Fig. 5.2. In Release 14 the unicast transmission schemes have not been enhanced or modified for the support of V2X traffic. On the other hand, the Multimedia Broadcast Multicast Service (MBMS) protocols have been evolved, with the introduction of a new logical function, namely V2X Application Server (AS), for network-related action required for V2X, similar to Group Communication Service Application Server as shown in Fig. 5.3.

As mentioned above, Release 14 C-V2X enables vehicles to directly exchange data over the so-called sidelink radio interface. At PHY layer, they both foresee the

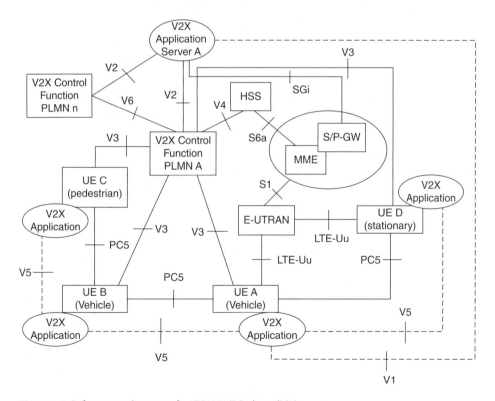

Figure 5.2 Reference architecture for LTE V2X (PC5-based) [5].

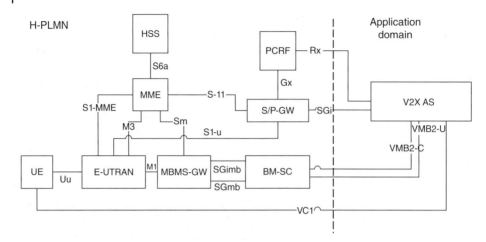

Figure 5.3 Reference architecture for LTE V2X (eMBMS-based) [5].

transmission on the unlicensed 5.9 GHz ITS band, employing either 10 MHz or 20 MHz wide channels. Moreover, they adopt the Single-Carrier Frequency Division Multiple Access (SC-FDMA) transmission scheme, with two extra Demodulation Reference Signals being inserted within the frame pattern; this warrants an improved channel estimation capability at the receiver, even under significant Doppler shifts. Specifically, whereas *Mode 3* still relies on the cellular network to perform radio resource assignment, Mode 4 does not, therefore appearing as the ideal candidate for satefy applications, required to flawlessly operate in out-of-coverage scenarios too. In Modes 3 and 4, every data message is paired with its Sidelink Control Information (SCI), transmitted within the same subframe and conveying crucial information for the successful decoding of the data message at the receiver side. Whereas the size of a subchannel is not standardized, the SCI always occupies two consecutive Resource Blocks (RBs).

5.2.1.2 Release 15 – *First Taste of 5G*
The first 5G standard was agreed in December 2017, with the Non-Standalone (NSA) deployment, where 5G New Radio operates on top of LTE core. Release 15 was finalised in June 2018 with the introduction of Standalone (SA) deployment and the introduction of the 5G core (5GC). In Phase 2 of V2X development within 3GPP, four more use cases are being addressed as part of Rel.15. These include Vehicles Platooning, Extended Sensors, Advanced Driving and Remote Driving. The aim of Rel.15 is to be backward-compatible to Rel.14 V2X for the delivery of safety messages. The work for this release is still in progress and several items have been identified to be studied, which span from physical layer enhancements for 64QAM and carrier aggregation for up to 8 PC5 carriers to modifications for shorter TTI and other enhancements to reduce the maximum time between packet arrival at Layer 1 and resource selecting for transmission.

5.2.1.3 Release 16 – *Fully-Fledged 5G*
The fully-fledged 5G experience, especially the Ultra-Reliable Low-Latency Communication (URLLC) service types, with both 5G New Radio (5G-NR) and 5G Core

(5GC), are expected to be available with the advent of Rel. 16. Particularly, the 5G-NR is a step beyond the LTE-A Pro that allows operations from low to very high bands (0.4–100GHz), scalable OFDM numerology with scaling of sub-carrier spacing, new channel coding schemes with polar and LDPC codes, ultra-wide bandwidth with 100MHz below 6GHz and more than 400MHz in >6GHz.

Release 16 is still under development, with several study items being considered for investigation as documented in the 3GPP report "Study on evaluation methodology of new V2X use cases for LTE and NR". The motivation is to:

- establish the evaluation methodology to be used in evaluating technical solutions to support the full set of 5G V2X use cases;
- further improve the evaluation methodology of LTE V2X and finalize it for NR V2X;
- investigate the channel model for sidelink in spectrum >6GHz; and
- identify the regulatory requirements and design considerations of potential operation of direct communications between vehicles in spectrum allocated to ITS beyond 6GHz in different regions, considering at least 63–64GHz (allocated for ITS in Europe) and 76–81GHz depending on regulatory decision.

Why 1 ms Latency is Needed? One millisecond latency have been voiced several times and is one of the marketing targets for 5G, quoted as a requirement for CAVs. Even though none of the service referenced in [6] has such a low latency requirement (the lowest being 3 ms for emergency trajectory alignment and the rest are above 10 ms), having 1 ms latency gains in scheduling and hence being able to support a larger number of vehicle with latency guarantee As an illustrative example, we assume a V2X control message with 1500 bits payload fits into a single TTI and there are a maximum of 2 retransmissions to achieve a reliability of 99.9% as requested by most V2X services. In LTE systems, Hybrid Automatic Repeat Request (HARQ) is employed, and its timing is illustrated in Figure 5.4. One HARQ cycle accounts for 8 TTIs. We also consider the new frame structure for 5G with 250 μs TTI length (M=14) and 60 kHz sub-carrier spacing. The question states "How many transmissions of safety messages can be supported with a latency guarantee of 20 ms?" With the current LTE-V2X system (Rel.14), one message per TTI of length 1 ms is transmitted. Due to the frame structure, and retransmission

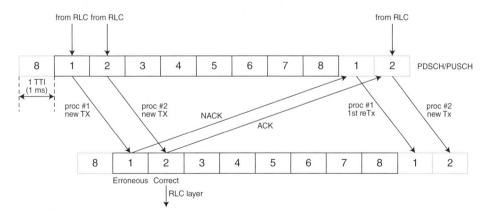

Figure 5.4 Hybrid automatic repeat request (HARQ) process.

scheme (HARQ), it can guarantee 17 ms latency, leaving 3 ms margin corresponding to 3 TTIs. Therefore, with LTE-V2X it is possible to support (1+3) = 4 safety messages with latency guarantee of 20 ms. On the other hand, in 5G and with the assumption of 5G-NR, TTIs of 250 μs, the latency guarantee of 1.5 ms leaves a margin of 18.5 ms margin which corresponds to 74 TTIs. Thus, a total of (1+74) = 75 safety messages can be supported with a latency guarantee of 20 ms. Hence, 5G can support almost 19× more vehicles with guaranteed latency within the same bandwidth.

5.2.2 Dedicated Short Range Communication (DSRC)

As mentioned in the introduction, DSRC is one of the two primary candidate technologies for vehicular communications, along with C-V2X. The access mechanism is based on the IEEE 802.11p standard, as it is standardised within IEEE 802.11-2016 Outside of the Context of a BSS (OCB) scheme [7]. IEEE 802.11p operates in distributed manner, adopting carrier sensing multiple access with collision avoidance (CSMA/CA). It does not require fine synchronization and provides the required reliability but suffers from poor performance when traffic load and number of users increase. Several studies have been conducted in analysing its performance, with an analytical model presented in [8].

The operating band is defined at 5.9GHz with 70 MHz allocated for ITS communications. The band is subdivided into 7 channels, each one of 10 MHz: one channel is used as control channel (CCH), whereas the remaining channels are for service (SCH). IEEE 802.11p is based on 64 OFDM sub-carriers, of which 52 are used for transmission (48 data sub-carriers and 4 pilots). Eight possible combinations of modulation and coding scheme (MCS) are provided; depending on the adopted MCS, the raw data rate varies between 3 and 27 Mb/s.

In Europe, the reference architecture for Cooperative ITS (C-ITS) from ETSI [9] specifies the communication protocols, e.g., Geo-Routing protocols, Basic Transport Protocol (BTP) as transport of ITS-specific traffic, and the ITS-G5, based on IEEE 802.11p, as the access layer for DSRC. On the other side of the Atlantic, the US primarily use IEEE WAVE architecture for DSRC. IEEE WAVE is a combination of IEEE 1609 family, IEEE 802.11-2016 and SAE J2735 standards [10]. DSRC-based solutions have been also tested in real systems and several automotive manufacturers plan to include it in future models.

With IEEE 802.11p being standardised also a decade now, IEEE recently formed a new study group, the 802.11 NGV, to study the Next Generation V2X. Its main aim is to focus on evolutions of the existing IEEE V2X systems that will support full backward compatibility and interoperability with existing deployed implementation based on IEEE 802.11-2016 (e.g. IEEE WAVE and ETSI ITS-G5). It will further investigate new use cases and frequency bands taking into account the regulatory frameworks across different regions.

5.2.2.1 Co-Existence
Given the fact that C-V2X and DSRC use different physical layers and medium access control protocols, the operation of the two technologies in the 5.9 GHz band and in the same geographic area without an agreed coexistence solution would result in mutually harmful co-channel interference. 5G Automotive Association (5GAA) has proposed

Figure 5.5 5GAA Co-Existence Options.

framework for co-existence of DSRC (IEEE 802.11p) and C-V2X. The proposal has three options (Figure 5.5), which could be seen evolutionary steps as technologies mature and shared spectrum could be used with reduced likelihood of harmful co-channel interference. It assumes that only 30MHz (5875MHz–5905MHz), out of totally 70 allocated for ITS, are used. However, one could extrapolate the solutions to 70MHz, with various options for more shared or more dedicated spectrum.

- Option A: At the early stages of V2X deployment, the two technologies[3] should each have a *safe harbour* of 10MHz at the two ends of the band, for dedicated usage. The remaining 10MHz should act as a guard to mitigate co-channel interference.
- Option B: Once the development and deployment of the two technologies mature, it could be possible to share the middle 10MHz channel, which was left unused in Option A. Employing techniques such as detect-and-vacate or listen-before-talk (LBT) that is used in LTE Assisted Access (LAA), the channel could be used in fair manner by the two technologies.
- Option C: Finally, the whole band could be used in an extended detect-and-vacate scheme for co-existence of the two V2X technologies. However, there would be some sort of prioritization on specific channels, with primary and secondary users for the two technologies.

5.2.3 Advanced Technologies

Apart from the advances in 5G-NR and 5GC, other advanced technologies, namely *Mobile Edge Computing* (MEC) and *Network Slicing*, are promoted as key enablers for URLLC and particularly V2X services.

5.2.3.1 Multi-Access Edge Computing

Mobile Edge Computing, lately re-branded by ETSI as Multi-Access Edge Computing, refers to a broad set of techniques designed to move computing and storage out of the remote cloud (public or private) and closer to the source of data. MEC is acknowledged as one of the key technology enablers for the emerging 5G use cases, especially the ones with stringent latency and throughput requirements. MEC plays an essential role in the telco business transformation, where the telecommunication networks are turning into versatile service platforms for industry verticals and other specific customer segments. It

---

3 Tech A and Tech B could be either C-V2X or DSRC, respectively

is a multi-vendor environment that is accessible by all multiple access technologies, not only 5G. Therefore, such a MEC platform could also bridge cellular and DSRC for some V2X services. MEC is confused with Fog computing and Cloudlets, but as we explain in [11], there are differences that are primarily related to the access and management of networking resources and the targeted use cases. The MEC platform is bringing the end services closer to the users, be it V2X applications or Virtual Network Functions (VNFs), providing a cost-effective hosting of compute and storage.

MEC is based on a virtualized platform, with an approach complementary to Network Function Virtualization (NFV). As can be seen in Figure 5.6, MEC framework can be grouped into system level, host level and network level entities. The mobile edge host is an entity that contains a mobile edge platform and a virtualization infrastructure which provides compute, storage, and network resources, for the purpose of running edge applications. The mobile edge platform is the collection of essential functionalities required to run mobile edge applications on a particular virtualization infrastructure and enable them to provide and consume mobile edge services. Mobile edge applications are processes that provide certain services at the edge of the network. These processes are typically executed on virtual machines, instantiated and maintained by the virtualization infrastructure. The mobile edge system level management includes the mobile edge orchestrator as its core component, which has an overview of the complete mobile edge system.

ETSI MEC framework can be used to provide a highly distributed cloud computing system with storage and processing both at the core and at the edge of the network,

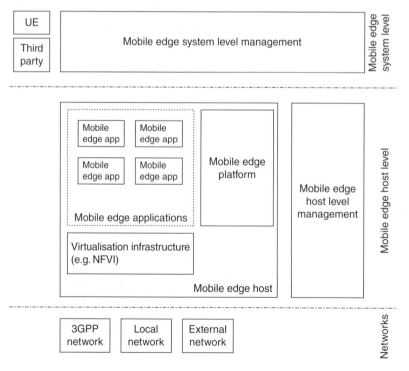

Figure 5.6 Mobile Edge Computing framework [12].

e.g., gNBs in an 5G system, close to connected vehicles. This can help significantly to reduce the response time of the system and enable a layer of abstraction in locality of the application from both the core network and applications provided over the Internet. The end user does not know where or who serves his query; the service may be provided by the edge host or the cloud. MEC applications may be deployed at a base station site and provide the roadside functionality. The MEC applications can receive local messages directly from applications in the vehicles and roadside equipment, analyse them and then propagate (with extremely low latency) hazard warnings and other latency-sensitive messages to other vehicles in the area. This enables a nearby car to receive data in a matter of milliseconds, allowing the driver to immediately react. The roadside MEC application will be able to inform adjacent MEC servers about the event(s) and, in so doing, enable these servers to propagate hazard warnings to cars that are close to the affected area. The roadside application will be able to send local information to the applications at the connected car cloud for further centralized processing and reporting.

5.2.3.2 Network Slicing

Network Slicing is the mechanism to logically separate networks, including provision of multiple control and data planes. There are different views among mobile operators and other stakeholders regarding the number of slices that would be available in future 5G systems. Some argue that the number would be in the tens to hundreds of slice profiles, that would match the QoS requirements for most of the use cases. On the other hand, there are others that suggest that a more granular split to 1000s of slices would be needed. It is generally challenging to specify a truly modular approach for slicing networks.

The variety of V2X applications with a distinct service requirement for QoS are a prime example where network slicing could be employed to deliver differentiated service. Each slice could be configured with different placement of VNFs, support of MEC or not, as shown in Fig. 5.7. For instance, an ambulance would typically require a higher priority, low-latency and reliable network slice to communicate with emergency services, while passengers in a car accessing infotainment services would access a high-bandwidth network slice. As the car moves from one cell to the other, the operator needs to deliver a continuous service with same QoS and 'slice parameters'. However, even in the same vehicle, different services could require different QoS and hence a slice. Sensor data used for predictive maintenance transmit infrequent small

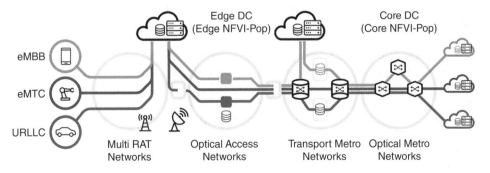

Figure 5.7 Network slicing example.

packets, while a camera used for See-Through service or for detecting road hazards and informing other vehicles in locality requires higher bandwidth and low latency.

5.3 Data Dissemination

Cooperative ITS system are expected to include several data providers and data ingestion services that reside on various entities within the system, and span from on-board sensor and actuators to cloud-based analytics. In order to support this complex ecosystem and enable those services to operate seamlessly and effectively, all entities should be capable to exchange information transparently, with low latency, reliably and securely. The mechanism to interconnect different systems and provide transparent application adaptations is commonly referred to as *middleware.*

There is no consensus on the definition of middleware, and in many cases it is subjective, depending on the perspective of those who define it. It generally consists of services and other resources located between the applications and the underlying data infrastructure, and acts as the "glue" to link different systems and data silos.

A formal definition of middleware can be found in RFC 2768 [13], which defines it as "a reusable, expandable set of services and functions that are commonly needed by many applications to function well in a networked environment". A simpler definition is given in [14] as the "software that connects computers and devices to other applications", which is later expanded to a more generic "software that lets other software interact with one another". Another middleware definition is to say that it acts as an intermediary. It is often used to support complicated and distributed applications such as those expected to operate in C-ITS.

There are many different types of middleware, as will be presented in the next sub-section, and each type will have its own specific definition depending on the challenge it aims to solve.

5.3.1 Context-Aware Middleware

A context is defined as a data source that can be sensed and used to characterize the situation of an entity [15]. Context-aware middleware is designed taking into account that data source, to optimise the operations and services provided by the middleware. Context-aware middleware is capable of wrapping, analyzing, and delivering physical world information to application services in a transparent way. As context definition is often domain-specific, there is no context-aware middleware solution to fit all systems. With respect to connected and autonomous vehicles, context may refer to the mobility of the vehicles, the applications and services offered by the vehicles and the infrastructure and lastly the resources available in the system. The requirements of a context-aware middleware are two-fold [16]. First, the functional, related to context acquisition, aggregation, context assessment, modelling, context dissemination, and privacy protection, and second operational, which consist of quality attributes such as performance, availability, usability, and extensibility, which should be satisfied by the system.

An example of context-aware middleware for mobile applications is CloudAware [17]. It considers both cloud and edge computing and supports automated context adaptation in the case of intermittent connectivity by linking the distribution features of mobile middleware with context-aware self-adaptation techniques. The middleware is based on Java Virtual Machine, providing an abstraction layer from the underlying operating system, to enable application offloading. In terms of context, it uses the device state, user information and the environment characteristics, performing connectivity predictions in order to support offloading towards the cloud or edge. A generic context adaptation service is introduced, in order to perform context prediction such as WiFi availability, bandwidth range and expected execution time of an offloaded task. CloudWare is based on Jadex [18], which provides a generic programming model based on active components that conceptually extend SCA (service component architecture). This allows for a transparent and unified access to the services and agents of the underlying (distributed) system. The Jadex middleware is responsible for all aspects of executing and administering active components.

The dynamics of the environment and diverse capabilities of devices make mobile application development and deployment challenging. A modular context-aware middleware is introduced in [19], which allows dynamic re-configuration of the mobile device with new peripherals, and hence new capabilities, at runtime. Applications are not required to be restarted in order to exploit the new information. The modularity allows developers to quickly and easily configure the middleware according to the needs of the deployed applications and the capabilities of the deployment platform. The challenge is to express and manage context information in a comprehensive and structured manner, which has lead to the development of Context Query Language and model. This allows for the development of application specific context, aggregate and analyse historical context information whilst maintaining the required privacy and traceability against the context usage.

One of the key challenges in mobile applications is energy efficiency, due to the battery constraints of the devices. Therefore, power consumption is expected to be a key source of contextual information. To that end, a power-aware middleware for mobile application is presented in [20]. The assumption is that processing on-board the mobile device is costly both in terms of time as well as energy, hence sensor data is transmitted towards the cloud in order to be processed. The solution allows for less complex user equipment with a simplified sensor data acquisition process. The data transmission to the cloud supports buffering to cope with intermittent connectivity. At the server side in the cloud, data processing is automated through intelligence and data is enhanced with spatio-temporal meta-data. Similarly, AURA [21] is a novel application-aware and user-interaction-aware energy optimization middleware. The objective of AURA is to optimize CPU and screen backlight energy consumption while maintaining a minimum acceptable level of user experience. AURA makes use of a Bayesian application classifier to dynamically classify applications based on user interaction. Once an application is classified, AURA utilizes Markov Decision Process (MDP) or Q-Learning-based power management algorithms to adjust processor frequency and screen backlight levels to reduce system energy consumption between user interaction events.

Connected and autonomous vehicle applications and services require both real-time and fault-tolerant interactions with the edge and cloud. A proposed cloud middleware for such use cases is presented in [22]. The architecture ensures fault-tolerance by automatically replicating virtual machines in data centres in a way that optimizes resources while assuring availability and responsiveness. Resource management is performed through an intelligent real-time publish/subscribe framework. However, the authors note that this is a "soft" real-time as hard real-time and safety-critical applications will not be hosted in the cloud. Nevertheless, this is can be debated with the advancements in fifth generation (5G) mobile communications for URLLC and the introduction of mobile edge computing. Further, fault tolerance based on redundancy is one of the fundamental principles for supporting high availability in distributed systems. This approach is followed by the work in [22], where a Local Fault Manager is instantiated at each physical host, and a replicated Global Fault Manager is used to manage a cluster of physical machines. The communication between Local and Global managers is performed though a high-availability middleware based on OMG DDS publish/subscribe messaging. In this framework, the Local managers publish the available resources in a timely manner to the Global manager, which subscribes to those events.

Lastly, middleware needs to be aware of the underlying hardware infrastructure, and in particular the available CPU and GPU resources. With the advancement of multi-core CPUs and multi-threaded applications in complex data centre configurations, the cost of thread migration on such a platform can be unpredictable and can lead to deteriorating service. Therefore, it is suggested in [23] that the middleware should avoid thread migration if possible. In situations that this is not possible, the authors propose MCFlow, a multi-core aware middleware that provides a very lightweight component model to facilitate system development. It also provides a mechanism for inter-component communication based on priority lanes with multiple receiving threads that handle incoming packets with different priorities. Using interface polymorphism to separate functional correctness from data copying and other performance constraints, MCFlow is able to configure and enforce these constrains independently but in a type-safe manner. The emergence of edge computing and spatially distributed applications has complicated the thread migration even more due to unavailability of context regarding the remote host. This is still an open issue identified in [11].

5.3.2 Heterogeneity and Interoperability

The diversity and heterogeneity of a C-ITS platform is evident both in the infrastructure and the services it shall provide. The middleware should provide seamless support for such a system by abstracting the differences between heterogeneous systems and exposing a uniform interface. IBM Altocumulus [24] is a cross-cloud middleware where its core manages the interactions with different cloud providers and exposes a RESTful API. The API facilitates the development of applications by homogenizing the clouds. A dashboard provides a user interface that uses the API to control and configure the underlying cloud infrastructure. A Mobile Cloud Middleware (MCM) presented in [25] addresses issues of interoperability across multiple clouds, dynamic allocation of cloud infrastructure and asynchronous delegation. The middleware abstracts the Web API of different/multiple cloud levels and provides a unique interface that responds (JSON-based) according to the cloud services requested (REST-based). Similar to Alticumulus, MCM hides the complexity of dealing with multiple cloud providers by abstracting the Web APIs from different clouds in a common operation level so that the service functionality of the middleware can be added based on combining different cloud services. In order to compensate for the diversity and randomness of processing delay at the cloud side, an asynchronous invocation is utilized through a push notification service. Adaptation is one of the challenges in mobile cloud computing, as different systems could talk "different languages". This can be tackled by the proposed middleware in [26]. The middleware adapts the provision of cloud web services by transforming SOAP messages to REST and XML format to JSON, and optimize the results by extracting relevant information, and improves the availability by caching.

Interoperability is commonly tackled by employing standards that service and equipment providers should comply. Such an approach is followed in cooperative intelligent transport, with standards being issued by ETSI and OneM2M. The ETSI C-ITS architecture describes the methods and service for cooperative vehicular applications, spanning from definitions of applications and network protocols to security and management. OneM2M standards have been employed in OneTrasnport[4] for an open, standardised, marketplace solution that can enable multimodal transport information to be published by data owners (e.g. transport authorities and new third parties) and accessed nationally by transport authorities, application developers and others. The 5G core network will be based on what is called "Service-Based Architecture" (SBA), centred around services that can register themselves and subscribe to other services. This enables a more flexible development of new services, as it becomes possible to connect to other components without introducing specific new interfaces. The new system architecture is specified in 3GPP technical specification 23.501 [27].

As discussed in [11], there is a paradigm shift towards containers and functions as a service. Container service has become mainstream and aims to provide portability without requiring software installations by bundling the application and its dependencies, including software libraries and a local filesystem. Containers' benefit is that they have smaller footprint compared to virtual machines, and hence are well suited for resource-constrained edge devices. Functions, on the other hand, could be part of an application and embrace the concept of micro-service architecture. Applications, then, can be deployed as sets of independent or interdependent functions, where each function is run in response to a given event generated from outside or inside the application. These two worlds were recently merged in an open source framework, OpenFaaS,[5] where functions are orchestrated by underlying container engines such as Docker Swarm or Kubernetes. From a commercial perspective, Amazon Web Services (AWS) introduced Greengrass, a software that lets users run local compute, messaging, data caching, and sync capabilities for connected devices in a secure way. AWS Greengrass can run AWS Lambda Functions at the edge, hence providing a common API between edge and core cloud offering. Similarly, IBM offers edge analytics, which enables the execution of Watson IoT Platform Analytics on a gateway device. In parallel, IBM's serverless offering is based on Apache OpenWhisk. LEON [28] allows Local Execution of OpenWhisk actions with Node-Red, which extends IBM's edge analytics with FaaS. Both AWS Greengrass and LEON are bound to underlying services from AWS or IBM, respectively. Each platform offers disparate sets of functionality, different user interfaces, and different runtime environments - even when working within the same language, which lead to potential vendor lock-in. However, cloud provider abstraction

---

4 http://www.interdigital.com/solution/onetransport
5 https://blog.alexellis.io/introducing-functions-as-a-service/

libraries such as libcloud,[6] fog,[7] and jclouds[8] provide unified APIs for accessing different vendors' cloud products.

5.3.3 Higher Layer Communication Protocols

Lastly, as mentioned earlier, vehicular communications can be divided to (a) on-board, (b) off-board local (vehicle-to-vehicle and vehicle-to-infrastructure) and (c) off-board remote (vehicle-to-network) connections. An abundance of sensors and actuators on-board a vehicle are connected to electronic control units (ECUs) through mainly wired networks such as CAN and FlexRay. Lately, camera and LiDAR sensors use high-capacity Ethernet links to support the high amount of generated data. These networks use either proprietary protocols or standard TCP/IP socket. Off-board local network use primarily vehicular-specific protocols such as those described in the ETSI C-ITS architecture. They have been been developed for safety-related applications disseminating information through single-hop broadcast messages. Connecting a vehicle to the cloud is governed primarily by the Internet protocol stack (TCP/IP). Web services are the dominant means of offering a service in the cloud, and as described previously this could potentially have variations to the data representation protocol (JSON vs. XML) used, which would require certain adaptation layers. The Internet of Things (IoT) paradigm, which also has been incorporated to vehicular networks, is governed by Message Queue Telemetry Transport (MQTT) protocol. Large cloud providers such as Amazon, IBM and Microsoft use MQTT as means to connect the end devices to their IoT hubs and aggregation points. MQTT is an ISO standard (ISO/IEC PRF 20922) publish-subscribe-based "lightweight" messaging protocol for use on top of the TCP/IP protocol. It provides three Quality of Service levels that relate to the message delivery guarantees. The first level, QoS-0, is the minimal level, providing a best-effort delivery without explicit acknowledgement other than the underlying TCP. This level is suggested to be used for stable connections between the counterparts, mainly over wired links. The second level, QoS-1, guarantees that the message will be delivered at least once. The receiver will send a PUBACK message back to the sender for acknowledgement of correct reception. QoS-1 is to be used when every message needs to be delivered and there is a mechanism to handle duplicates. Finally, QoS-2 is the highest level of service, guaranteeing only one reception of the message. It is the safest but also the slowest quality of service level, reducing the potential overhead from duplicates that exist in QoS-1. The guarantee is provided by two flows there and back between sender and receiver. Benchmarking of MQTT with different configurations is performed to evaluate its capabilities in nominal scenarios using an open source tool.[9] As expected, QoS-2 flows exhibit higher delay and lower throughput due to overhead of acknowledgement flows. Interestingly, the packet size does not impact the performance significantly as shown in Figure 5.8. However, the increase in number of clients publishing to a single broker has a linear impact to both delay and throughput. The second set of benchmarking experiments uses HiveMQ[10] to evaluate two common uses

---

6 http://libcloud.apache.org
7 http://fog.io
8 www.jclouds.org
9 https://github.com/krylovsk/mqtt-benchmark
10 http://www.hivemq.com/benchmarks/

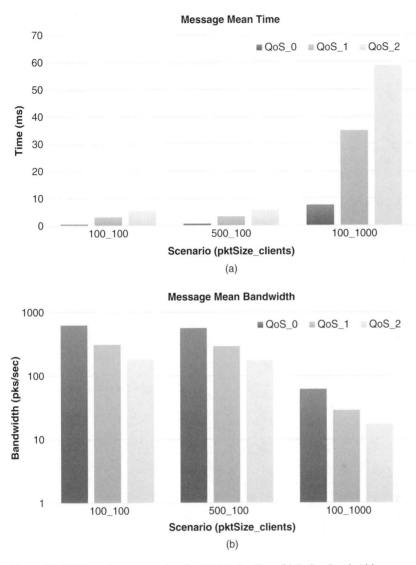

Figure 5.8 MQTT performance on localhost (a) QoS vs Time (b) QoS vs Bandwidth.

of MQTT, the telemetry and fan-out, with results in Figure 5.9. In the telemetry test, multiple clients publish to a limited number of subscribers. This is useful for CAV when vehicles want to report their location and other information to a centralized entity. The telemetry tests showed that HiveMQ handles more than 60,000 messages/second with minimal resource consumption (RAM < 3GB) with linear increasing throughput up to 15 MB/s for each, incoming and outgoing traffic. The fan-out test is the reverse of telemetry. A single publisher broadcasts to a large number of subscribers, which is common in CAVs when an update has to be disseminated to a large number of vehicles. HiveMQ served up to 150,000 messages/second with 33 MB/s to subscribers with

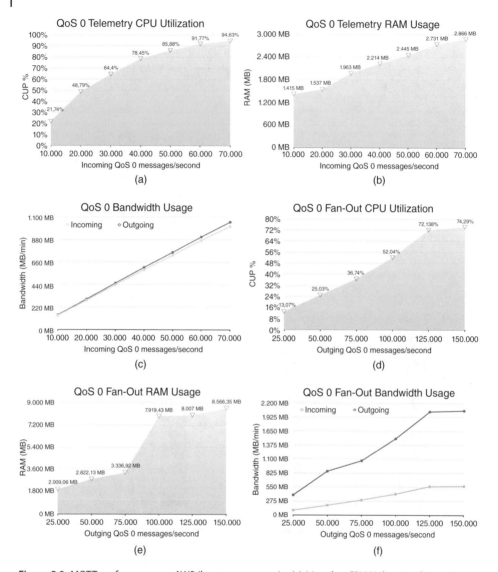

Figure 5.9 MQTT performance on AWS (host on same region) (a) Load vs CPU Utilization (b) Load vs RAM Utilization (c) Load vs Bandwidth (d) Load vs CPU Utilization (e) Load vs RAM Utilization (f) Load vs Bandwidth.

medium CPU utilization and used RAM below 9 GB. Due to disk persistence, HiveMQ meet all QoS 1 and 2 guarantees in the fan-out tests at the cost of a lower message rate per second.

There are several alternatives to MQTT for IoT networking. A comprehensive qualitative review of the most common ones is available in [29]. Therein, Data Distribution Service for Real-Time Systems (DDS) is also analysed. DDS is a publish/subscribe protocol without the need of a broker. It uses a common data bus over which the communication is performed, which makes it challenging for off-board communications. However,

DDS could be a suitable candidate for on-board communications in CAV systems, where sensors publish their readings to multiple ECUs and actuators subscribe to real-time commands, or for services living at the edge and wanting to communicate with each other. Lastly, as discussed in Section 5.3.1, context is used by middleware to optimize its operation for several aspects. A new Meta-Data Protocol (MDP) is being proposed by 5G Innovation Centre [30] that enables secure and efficient exchange of contextual information between network and users. The user can selectively (but securely) share information with the network and other third party stakeholders that the user enables in order to improve the users experience directly or as a result of empowering the access network to better support the user request with priority information from the user.

5.4 Towards Connected Autonomous Driving

The evolution of V2X technologies as illustrated in Figure 5.10 has enabled several innovative applications that in the long run will turn CAVs to full autonomy with accident free-driving. The roadmap to enable that is depicted in Figure 5.11. Development of Phases 1 & 2 is almost complete and it is now at the phase of mass deployment in the field. The focus of Research is on Phases 3 & 4, with early stages of autonomous vehicles hitting the roads across the globe. The 5G Infrastructure Public Private Partnership (5GPPP) has recently published a study on the deployment strategies for V2X services, with the ultimate goal the deployment of 5G infrastructure. One of the main findings of that work is that key parameters influencing the profit of CAV services are the OPEX/CAPEX costs, which also depend on the percentage of usage of the infrastructure for ITS services.

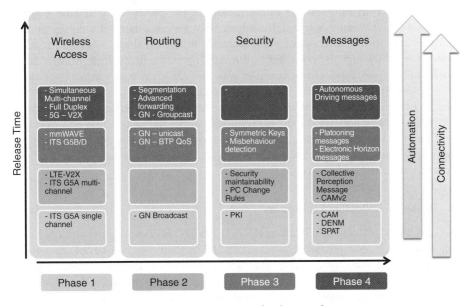

Figure 5.10 Car-to-car communication consortium technology roadmap.

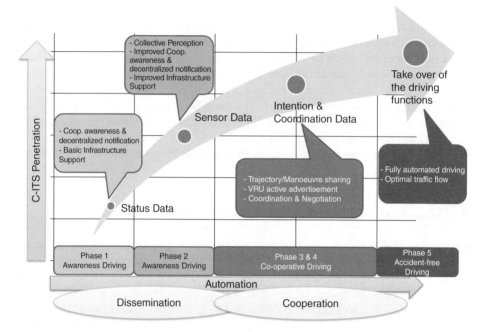

Figure 5.11 Car-to-car communication consortium application roadmap.

5.4.1 Phase 1 – Awareness Driving Applications

The first set of C-ITS applications, also known as Day-1 applications, are based around raising awareness around the vehicles. This is done through the Cooperative Awareness Messages (CAM) and Decentralized Environment Notification Messages (DENM) that are broadcasted from vehicles and infrastructure either periodically or after events. Vehicles share their dynamics and status information so that others would infer potential harmful conditions, such as collision. Roadside units also share event information generated internally such as traffic light information, or forward vehicle notifications. Deployment of these applications is under way and OEMs are preparing their fleet. These applications will be initially served by DSRC systems as the requirements for latency and data rates are low. Even though the penetration is expected to be low, it is expected to have a positive impact both on safety and on traffic efficiency. Example applications for this set include intersection collision warning, traffic jam warning, motorcycle approach warning, pre/post-crash warning, in-vehicle signage, traffic light information and roadworks warning.

5.4.2 Phase 2 – Collective Perception

The second phase of C-ITS applications is based around collaborative perception. Vehicles share abstract descriptions of objects detected by their own sensors or infrastructure sensors in order to create an improved awareness around them even though when C-ITS penetration is low. Such services include overtaking warning, vulnerable road users (VRU) warning, cooperative ACC and special vehicle prioritization. The communication system for that could be either DSRC or cellular. For example, the 5G-Safe project is exploring methods to disseminate 3D surround views through 5G communications. Further, sensor data from cameras can be shared to identify VRU. In the long run, and as C-V2X technology becomes more prevalent, there is a potential to use direct communications with mobile phones carried by pedestrians to improve awareness and safety.

5.4.3   Phase 3/4 – Trajectory/Manoeuvre Sharing

As the level of automation increases and the driver variability is removed from the equation, vehicles are able to be more confident of their future trajectories. Sharing their intended manoeuvres and trajectories with other automated vehicles can significantly help with planning and improving vehicle control. For example, the MAVEN project will develop infrastructure-assisted algorithms for the management of automated vehicles, which connect and extend vehicle systems for trajectory and manoeuvre planning. They are testing applications beyond the state-of-the-art of ADAS and C-ITS services like GLOSA, by adding cooperative platoon organization and signal plan negotiation to adaptive traffic light control algorithms. Simultaneously these algorithms will yield substantial better utilization of infrastructure capacity and reduction of vehicle delay and emission, while ensuring traffic safety. Similarly, TransAID develops solutions for infrastructure-assisted traffic management procedures, protocols and guidelines for smooth coexistence between automated, connected and conventional vehicles, especially at Transition Areas. Currently, these projects are using DSRC communications for prototyping the solution.

5.4.4   Phase 5 – Full Autonomy

Level five autonomy will take over the driving tasks and, as many evangelise, deliver an accident-free and optimal traffic flow travel experience. However, recent incidents involving autonomous vehicles have sparked negative press and underscored the importance of public safety in self-driving cars. These events have also highlighted the challenges facing the industry to develop autonomous driving systems that can guarantee a safety performance above that of human drivers. One of the key observations is that current autonomous vehicles are not connected. Cooperation among vehicles, autonomous or not, will greatly improve the awareness as it is shown with the initial connected vehicle phase. Further, it could be possible to remotely operate autonomous vehicles from remote facilities to assess live video feeds and vehicle diagnostics from the autonomous vehicles, and take over driving control virtually. However, this would require the reliability and low latency that only 5G networks could provide.

5.5   Conclusions

Vehicles and roads are starting to be connected and gradually moving towards fully autonomous vehicles and truly intelligent road infrastructure. There are plenty of challenges ahead, although some of these are not related to technology. The two main communication systems, DSRC and C-V2X, are evolving and potentially converging towards 5G, with advances both in the radio access technologies as well as in the network architecture. Although the key technology enablers for 5G vehicle-to-anything (V2X) communication are, meanwhile, well understood in the cellular community and standardization of 3GPP Rel. 16 V2X is about to start, there is still lack on insights into the required roll-out investments, business models and expected profit for future Intelligent Transportation Services (ITS).

References

1 Z. MacHardy, A. Khan, K. Obana, and S. Iwashina. V2X Access technologies: regulation, research, and remaining challenges. *IEEE Communications Surveys Tutorials*, 20: 1858–1877, 2018. doi: 10.1109/COMST.2018.2808444.

2 A. Bazzi, B. M. Masini, A. Zanella, and I. Thibault. On the performance of IEEE 802.11p and LTE-V2V for the cooperative awareness of connected vehicles. *IEEE Transactions on Vehicular Technology*, 66(11): 10419–10432, 2017. ISSN 0018-9545. doi: 10.1109/TVT.2017.2750803.

3 Z. H. Mir and F. Filali. On the performance comparison between IEEE 802.11p and LTE-based vehicular networks. In *2014 IEEE 79th Vehicular Technology Conference (VTC Spring)*, pages 1–5, May 2014. doi: 10.1109/VTCSpring.2014.7023017.

4 A. Vinel. 3GPP LTE versus IEEE 802.11p/WAVE: which technology is able to support cooperative vehicular safety applications? *IEEE Wireless Communications Letters*, 1(2):125–128, April 2012. ISSN 2162-2337. doi: 10.1109/WCL.2012.022012.120073.

5 3GPP. Study on architecture enhancements for LTE support of V2X services. TR 23.785, 3GPP, 2016.

6 3GPP. Service requirements for enhanced V2X scenarios. TS 22.186, 3GPP, 2015.

7 IEEE. IEEE Standard for Information Technology – Telecommunications and information exchange between systems Local and metropolitan area networks–Specific requirements Part 11: Wireless LAN Medium Access Control (MAC) and Physical Layer (PHY) Specifications. *IEEE Std 802.11-2016*, 2016.

8 C. Han, M. Dianati, R. Tafazolli, R. Kernchen, and X. Shen. Analytical study of the IEEE 802.11p MAC sublayer in vehicular networks. *IEEE Transactions on Intelligent Transportation Systems*, 13(2):873–886, June 2012. ISSN 1524-9050. doi: 10.1109/TITS.2012.2183366.

9 ETSI. Intelligent Transport Systems (ITS) – Communications Architecture. ETSI EN 302 665, ETSI, 2010.

10 IEEE. IEEE Guide for Wireless Access in Vehicular Environments (WAVE) – Architecture. 1609.0-2013, IEEE, 2013.

11 Lee Gillam, Konstantinos Katsaros, Mehrdad Dianati, and Alexandros Mouzakitis. Exploring edges for connected and autonomous driving. In *CCSNA: Cloud Computing Systems, Networks, and Applications, INFOCOMM Workshop*, 2018.

12 ETSI. Mobile Edge Computing (MEC) – Framework and Reference Architecture. ETSI GS MEC 003, ETSI, 2016.

13 B Aiken, J Strassner, B Carpenter, I Foster, C Lynch, J Mambretti, R Moore, and B Teitelbaum. Network policy and services: A report of a workshop on middleware. RFC 2768, IETF, 2000.

14 Cloud Computing Glossary – Cloud Middleware. http://apprenda.com/library/glossary/definition-cloud-middleware/. Accessed: 24 July 2017.

15 O. Yurur, C. H. Liu, and W. Moreno. A survey of context-aware middleware designs for human activity recognition. *IEEE Communications Magazine*, 52(6):24–31, June 2014. ISSN 0163-6804. doi: 10.1109/MCOM.2014.6829941.

16 Hamed Vahdat-Nejad. *Context-Aware Middleware: A Review*, pages 83–96. Springer New York, 2014. ISBN 978-1-4939-1887-4. doi: 10.1007/978-1-4939-1887-4_6.

17 G. Orsini, D. Bade, and W. Lamersdorf. CloudAware: a context-adaptive middleware for mobile edge and cloud computing applications. In *2016 IEEE 1st International Workshops on Foundations and Applications of Self* Systems (FAS*W)*, pages 216–221, Sept 2016. doi: 10.1109/FAS-W.2016.54.

18 Alexander Pokahr, Lars Braubach, and Kai Jander. Jadex: a generic programming model and one-stop-shop middleware for distributed systems. *PIK-Praxis der Informationsverarbeitung und Kommunikation*, 36(2):149–150, 2013.

19 Nearchos Paspallis and George A. Papadopoulos. A pluggable middleware architecture for developing context-aware mobile applications. *Personal and Ubiquitous Computing*, 18(5):1099–1116, 2014. ISSN 1617-4909. doi: 10.1007/s00779-013-0722-7.

20 R. Pérez-Torres and C. Torres-Huitzil. A power-aware middleware for location amp; context aware mobile apps with cloud computing interaction. In *2012 World Congress on Information and Communication Technologies*, pages 691–696, Oct 2012. doi: 10.1109/WICT.2012.6409164.

21 Sudeep Pasricha, Brad K. Donohoo, and Chris Ohlsen. A middleware framework for application-aware and user-specific energy optimization in smart mobile devices. *Pervasive and Mobile Computing*, 20(C):47–63, 2015. ISSN 1574-1192. doi: 10.1016/j.pmcj.2015.01.004.

22 Kyoungho An, Shashank Shekhar, Faruk Caglar, Aniruddha Gokhale, and Shivakumar Sastry. A cloud middleware for assuring performance and high availability of soft real-time applications. *Journal of System Architecture*, 60(9):757–769, 2014. ISSN 1383-7621. doi: 10.1016/j.sysarc.2014.01.009.

23 Huang-Ming Huang, Christopher Gill, and Chenyang Lu. MCFlow: a real-time multi-core aware middleware for dependent task graphs. In *Proceedings of the 2012 IEEE International Conference on Embedded and Real-Time Computing Systems and Applications*, RTCSA '12, pages 104–113, 2012. doi: 10.1109/RTCSA.2012.30.

24 Ajith Ranabahu and Michael Maximilien. A best practice model for cloud middleware systems. In *24th ACM SIGPLAN International Conference on Object-Oriented Programming, Systems, Languages, and Applications (OOPSLA)*, pages 41–51, 2009.

25 Huber Flores and Satish Narayana Srirama. Mobile Cloud Middleware. *Journal of Systems and Software*, 92:82–94, 2014. doi: http://dx.doi.org/10.1016/j.jss.2013.09.012.

26 K. Akherfi, H. Harroud, and M. Gerndt. A mobile cloud middleware to support mobility and cloud interoperability. In *2014 International Conference on Multimedia Computing and Systems (ICMCS)*, pages 1189–1194, April 2014. doi: 10.1109/ICMCS.2014.6911331.

27 3GPP. System Architecture for the 5G System – stage 2 (Rel.15). TS 23.501, 3GPP, 2017.

28 Alex Glikson, Stefan Nastic, and Schahram Dustdar. Deviceless edge computing: extending serverless computing to the edge of the network. In *Proceedings of the 10th ACM International Systems and Storage Conference*, SYSTOR '17, pages 28:1–28:1. ACM, 2017. ISBN 978-1-4503-5035-8. doi: 10.1145/3078468.3078497.

29 Andrew Foster. Messaging Technologies for the Industrial Internet and the Internet of Things. White paper, PRISMATECH, 2015.

30 5GIC. The Flat Distributed Cloud (FDC) 5G Architecture Revolution. White Paper, ICS-5GIC, 2016.

6

State-of-the-Art of Sparse Code Multiple Access for Connected Autonomous Vehicle Application

Yi Lu, Chong Han, Carsten Maple, Mehrdad Dianati and Alex Mouzakitis

Future applications for Connected Autonomous Vehicles (CAVs), such as traffic management, enhanced safety, crash response, integrated navigation, infotainment etc., will require robust communications and networking capabilities. With the emergence of the fifth-generation (5G) wireless communication technology, it promises a new view to support CAVs communications systems with high reliability and low latency. Considering the further application of CAVs with 5G wireless networks, new technologies have been investigated. One is multiple access technology called Sparse Code Multiple Access (SCMA). This chapter aims to investigate the fundamental principles of SCMA technique, summarize the state-of-the-art of SCMA technology and its application, and point out how the current research work can contribute to the quality of experience for drivers and passengers. In addition, this work also identifies the challenges and research gaps for the provision of massive connections with low latency and high reliability in the 5G communication system for CAVs.

6.1 Introduction

In the last few years, governments, universities and industries from many countries have committed millions of pounds to exploring Connected and Autonomous Vehicles (CAVs), and the research and development of CAVs are emerging. Although the realization of CAVs is still at its early stage, the intensity of investigation of CAVs ranging from theoretical to experimental are being undertaken. The full autonomous vehicle is described as a vehicle in which a driver is not necessary, and the connected vehicle technologies allow vehicles to talk to each other and to everything [1]. The combining of connected vehicles and autonomous vehicle allows safer, quicker and more efficient movement, which is achieved by establishing computer-driven vehicles mode. The main technologies to enable the CAVs include vehicle-to-vehicle (V2V) technology, vehicle-to-infrastructure (V2I) technology and vehicle-to-everything technology (V2X).

The V2I, V2V and V2X technologies are very promising to increase the safety and efficiency of the future CAVs networks. Among these, an important requirement for future V2X communication systems is that it should support massive connectivity with a large

Enabling 5G Communication Systems to Support Vertical Industries, First Edition.
Edited by Muhammad Ali Imran, Yusuf Abdulrahman Sambo and Qammer H. Abbasi.

amount of V2X devices [2]. Taking into account the fast-increasing number of intelligent vehicles on road (e.g., 400 vehicles per km²) [3] as well as mobile devices equipped with V2X applications such as devices for cyclists and pedestrians, and Roadside Units (RSUs), it becomes even more urgent to ensure that a large number of active users have access to the wireless medium simultaneously for the purpose of road safety enhancement. Moreover, offering stable connections for V2X devices is far more challenging than the provision of the same service for mobile handset users with low mobility.

A digital communications network can be considered as a transmission medium for a large number of user terminals to access. Commonly used access techniques assign fixed frequencies, time slots, or code sequences to individual transmitting users and these techniques are known as Frequency Division Multiple Access (FDMA), Time Division Multiple Access (TDMA), Code Division Multiple Access (CDMA) and Orthogonal Frequency Division Multiple Access (OFDMA), which are shown in Figure 6.1. FDMA is the method of dividing one channel or bandwidth into multiple individual bands, each band used by a single user. The best example of using FDMA is the cable television system. TDMA is another multiple access technology that divides a single channel or band into time slots. It is used in the 2G cellular systems such as Global System for Mobile Communication. There is no restriction on the time and frequency in CDMA; users are separated by code, not by time slot or frequency slot. CDMA is used in commercial cellular communications to better use the radio spectrum when compared with other technologies. OFDMA is the technique that used in Long-Term Evolution (LTE) cellular systems to accommodate multiple users in a given bandwidth. OFDMA is based on Orthogonal Frequency Division Multiplexing (OFDM), where the bandwidth is divided into a large number of smaller bandwidths that are mathematically orthogonal using fast

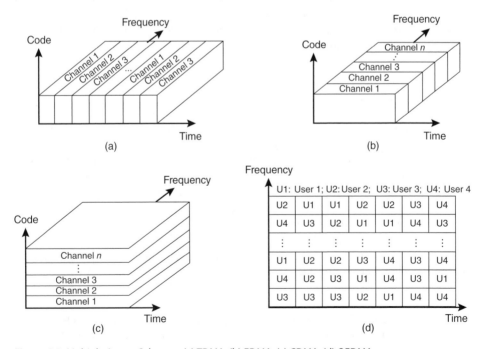

Figure 6.1 Multiple Access Schemes: (a) TDMA; (b) FDMA; (c) CDMA; (d) OFDMA.

Fourier transforms. It can be achieved by assigning subsets of sub-carriers to individual users. This allows simultaneous low data rate transmission from several users.

Nowadays, 5G technology is becoming the next frontier of innovation for the wireless systems. It will support the enhancement of mobile Internet, Internet of Things (IoTs) and future CAVs systems. The 5G technology aims to deliver high-quality service, ultra-capacity, ultra-high data rate, ultra-low latency, massive connectivity and ultra-low energy consumption. However, the multiple access technologies mentioned above cannot satisfy these requirements. Thus, new technologies have been proposed for 5G, such as SCMA [4] and Non-Orthogonal Multiple Access (NOMA) [5], to improve multiple access for massive active users simultaneously. Here, the focus will be on SCMA.

Since SCMA is proposed based on CDMA and its enhancement, a detailed background of CDMA is given before going into SCMA.

CDMA is a well-known multiple access technique in which the data symbols are spread out over orthogonal or near-orthogonal code sequences [6]. A CDMA receiver takes advantage of the orthogonality feature of the spreading sequences to extract the original data symbols carried over sequences or signatures. The sequence design is an important factor for a CDMA system in terms of the performance and the reception complexity. A CDMA encoder expands a Quadrature Amplitude Modulation (QAM) symbol to a sequence of complex symbols by using a given CDMA signature. The modulator in CDMA can be treated as a process in which a number of coded bits are mapped to a sequence of complex symbols.

From the principles of decoding CDMA, it can be seen that the longer the code is or the more codes the system has, the more complex the decoding becomes. Hence, in CDMA, there is another approach to help reduce the complexity on the receiver side, namely Low-density signature (LDS) [7]. LDS is a special approach to CDMA sequence design with a few numbers of non-zero elements within a larger signature length. In LDS, the spreading factor is the signature length. The low-density characteristic of LDS sequences makes it possible to reuse the message passing algorithm (MPA) in the complex domain for iteratively joint multi-user detection with near optimal performance.

Taking the above-mentioned techniques into account, SCMA is proposed as a multi-dimensional codebook-based non-orthogonal spreading technique that can be treated as a generalization of LDS. Similar to CDMA, the codebook design and pilot sequence design are important to SCMA's functional capability. In literature [8], the codebooks are usually assumed to be known by all entities in the network. An SCMA encoder includes a certain number of users each with some separate layers. Each user is usually assumed to have one layer for simplicity. SCMA supports an overloaded system where the number of non-orthogonal layers or users are larger than the spreading factor. At the receiver side, a decoder takes advantage of a near optimal MPA with an acceptable performance.

To be more specific, unlike LDS, SCMA directly maps incoming bits to multi-dimensional complex codewords selected from a predefined codebook set. Each layer or user has a specific SCMA codebook set. The overlaid layers or users are non-orthogonally superimposed on top of each other. SCMA replaces QAM modulation and LDS spreading with multi-dimensional codebooks. Thus, shaping gain of a multi-dimensional constellation is one of the main sources of the performance improvement in comparison to the simple repetition of QAM symbols in LDS.

In this chapter, the discussion of how state-of-art multiple access can contribute to the quality of experience for drivers and passengers in the CAV is presented. The focus of the work is on the new air interface access candidate technique for 5G, SCMA in two main areas. Firstly, the fundamental principle of the SCMA technique will be investigated. Secondly, the state-of-the-art of the SCMA technique is presented as regards three aspects: codebook design with regard to system reliability; decoding techniques with regard to decoding complexity; and other research work on system performance evaluation.

6.2 Sparse Code Multiple Access

SCMA, introduced in [4], is a frequency domain multiple access scheme which is a codebook-based non-orthogonal access mode that enables massive connectivity. It is designed with the following properties. Firstly, binary data bits are spread and encoded to a complex vector namely codeword; secondly, multiple access is achieved by generating multiple codebooks for each user; thirdly, codewords in the codebooks are sparse to give a moderate complexity decoding process; and finally, the system can be overloaded (e.g., the number of multiplexed layers can be more than the spreading factor).

SCMA codebook design can flexibly adapt to meet different system requirements by adjusting the SCMA parameters. From the point of view of CDMA, SCMA merges the QAM mapper block and the CDMA spreader to directly map a set of bits to a complex vector so called a "codeword". Hence, the spreading encoder in SCMA can be interpreted as a coding procedure from the binary domain to a multi-dimensional complex domain. An SCMA encoder is defined as a map from $\log_2 M$ bits to a K -dimensional complex codebook of size N. K-dimensional complex codewords of the codebook are sparse vectors with $N < K$ non-zero entries. All the codewords in the codebook contain 0 in the same $K - N$ dimensions. An SCMA encoder contains J separate layers, where $J = \binom{K}{N}$. As shown in Figure 6.2, the number of codebooks can be severely impacted by the length of codewords, K, and the number of non-zero entries, N, within the codewords.

The coding procedure from binary domain to a multi-dimensional complex domain is illustrated in Figure 6.3. The coding process can be deemed as merging the QAM mapper block and the CDMA spreader together, as in CDMA. Hence, it can be seen that, compared to the traditional CDMA, constellation shaping gain is the advantage of the multi-dimensional codeword modulation technique. Of course, designing complex multi-dimensional codeword sets is more challenging than a relatively simple CDMA signature design.

Each user transmits data using a specific codebook which consists of a set of codewords for different binary date interpretation. The example codebooks given in Figure 6.4 are generated when the length of the codeword (K) is 4 and the number non-zero entries in the codeword (N) is fixed at 2. When users have binary data to transmit, they are simply mapped to different non-orthogonal codewords which are overlaid over the same channel simultaneously. On the receiver side, data detection requires only channel information on the non-zero elements of the codeword. Hence, the complexity of decoding is largely reduced to achieve the desired acceptable detection performance.

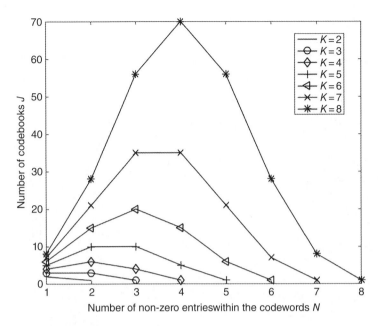

Figure 6.2 Number of codebooks as a function of *K* and *N*.

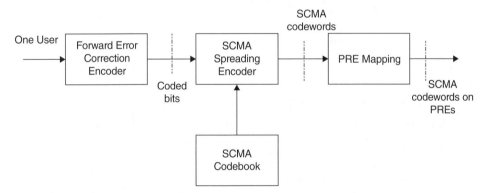

Figure 6.3 Encoding procedure in SCMA.

The characteristics of SCMA make it suitable for massive connection scenarios, especially for uplink connections, such key functions being, e.g. codebook design, decoding techniques, pilot sequences, collision avoidance mechanisms and blind detection technique. The codebook design and decoding schemes are important for the system complexity and will certainly affect the system performance. A lot of the current literature is focused on the investigation of these two functions. These will be reviewed in detail in Section 6.3. Here, pilot sequences, collision avoidance mechanisms and blind detection technique will be discussed.

Pilot sequences: In order to support uplink SCMA with contention-based access, radio resources are defined [9]. The basic resource for contention transmission is called a contention transmission unit (CTU), shown in Figure 6.5, which combines time,

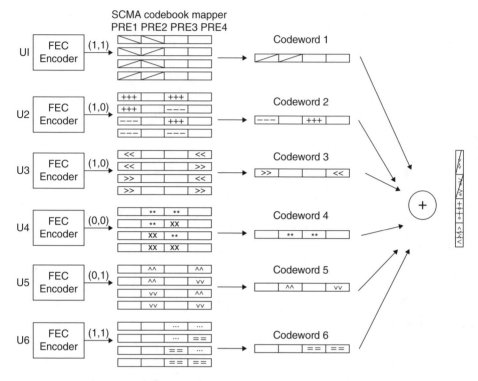

Figure 6.4 SCMA encoder ($K = 4$, $N = 2$ and $J = 6$).

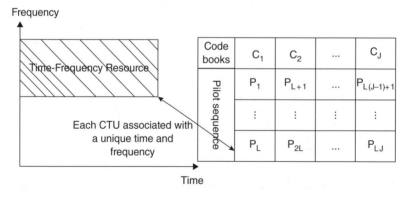

Figure 6.5 Contention transmission unit definition [9].

frequency, SCMA codebook and pilot sequence. Here, it is assumed that J unique codebooks are defined over time-frequency resource. For each codebook, L pilot sequences are associated with it. Thus, $L*J$ unique pilot sequences are defined. It can be seen that pilot sequences add another dimension to the available differentiated service. It largely extends the number of CTUs, which no doubt helps reduce the collision probabilities in contention-based multiple access systems, e.g., connected vehicles.

Collision avoidance mechanisms (back-off procedure): Although the number of available CTUs is largely extended (to $L*J$) thanks to the added dimension by pilot sequence, there is still possibility that multiple user equipments (UEs) can be mapped to the same CTU for data transmission. Hence, a collision can occur when more than one UE has data to transmit at the same time, selecting the same CTU. In the current SCMA scheme, collision resolution can employ the random back-off procedure. Each UE picks a random back-off time (in units of transmission time interval (TTI)) from a back-off window and retransmit on the predefined CTU as the original transmission.

Collision Avoidance Mechanisms (contention region selection): To avoid collisions caused by the same choice over CTUs, there are some mapping rules have been proposed. For instance, users can be mapped into different contention regions first which may be either predefined or derived following a specific criterion. Pre-defining contention regions requires the knowledge of users and has to be undertaken by a centralized control which is usually assumed absent in the contention-based access system. In addition, it involves too much signalling overhead to pre-allocate contention regions, while the simple UE-to-CTU mapping rule presented in [9] may not be effective in terms of utilization of radio resources.

Blind Detection: The blind detection technique for SCMA receiver design has been proposed in [10]. Thanks to this technique, SCMA becomes an enabler of grant-free multiple access since no approval is requested before selecting codebooks and data transmission attempts. The designed receiver consists of two key components: an active UE detector and joint data and active codebook detection (JMPA). As shown in Figure 6.6, the blind detection technique requires an up-to-date list of active UEs. The active UE detector block is to narrow down the list of potentially active UEs via the received signals. JMPA is to decode the active users' data without knowledge of active codebooks. JMPA has to test all the possible combinations of active UEs and its complexity becomes impractically high if too many potential UEs are added to the list. Hence, by updating the active UE detector with a priori existence probabilities for pilots, JMPA tries to assist the active UE detector to make the list as accurate as it can to reduce the complexity for decoding. With the predefined contention regions and CTU assignments to UEs, and the blind detection techniques, the dynamic uplink grant is no longer necessary for contention-based SCMA. Since uplink grants occupy

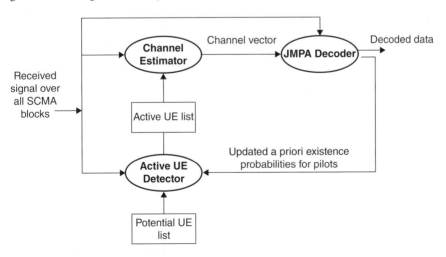

Figure 6.6 Blind detection receiver [10].

downlink control channel resources, elimination of grant signalling for traffic with uplink contention-based SCMA means that downlink signalling overhead can be reduced.

6.3 State-of-the-art

SCMA supports multiple UEs to share the same time-frequency resources through mapping to different codewords of SCMA codebooks. Massive connections can be achieved through SCMA due to the non-orthogonal property. All the features of the SCMA make it quite suitable for infrequent small packets of communications such as for machine-type communication (MTC) devices [11]. It is reasonable to consider the extended utilization of SCMA in vehicular networks where the event-driven message is similar to the traffic flows in MTC systems.

This section provides the state-of-the-art of SCMA with focus on performance analysis and enhancements. The aim of this literature review is to identify the gaps in the adoption of SCMA in vehicular communication systems.

6.3.1 Codebook Design

SCMA currently relies on a predefined codebook set in order to encode/decode data between binary domain and multi-dimensional complex domain. Codewords of the codebooks have to be sparse to reduce the complexity at the receiver. Codebook design in SCMA is complicated as multiple users or layers are multiplexed with different codebooks. Thus, designing an optimal SCMA codebook is still an open and challenging problem. Most of the current research on SCMA performance is about increasing the minimum Euclidean distance (e.g. [8]) of the mother constellation and paying attention to other parameters such as the energy diversity of the mother constellation and the layer specific operators for codebook generation [12]. Current research on this topic will be discussed in the following paragraphs.

The authors in [4] firstly defined the SCMA and proposed a sub-optimal systematic approach to design a set of codebooks. The SCMA and baseline LDS systems are evaluated through link-level simulations. In the given evaluations, the length of the codeword is set to 4, a non-zero value is 2, hence both the number of LDS signatures and SCMA codebook layers are 6. It means the overloading factor is up to 150%. When the rate of the turbo channel code is 3/4, the overall spectral efficiency is set to 2.25 bits per tone.

The authors in [8] propose another innovative codebook design by rotating constellation. The large minimum Euclidean distance of a multi-dimensional constellation ensures a good performance of SCMA systems. When the number of layers grows, 2 or more layers may collide over the same tone. Hence, a unitary rotation can be applied to the base constellation to control dimensional dependency and power variation of the constellation while maintaining the Euclidian distance profile unchanged. An example of a 16-point SCMA constellation applicable to codewords with 2 non-zero values is given in Figure 6.7. The idea is to construct an N-dimensional complex mother constellation from the Cartesian product of two N-dimensional real constellations. Each of the real constellations is constructed by the same method. One of the real constellations corresponds to the real part and the other one corresponds to the imaginary

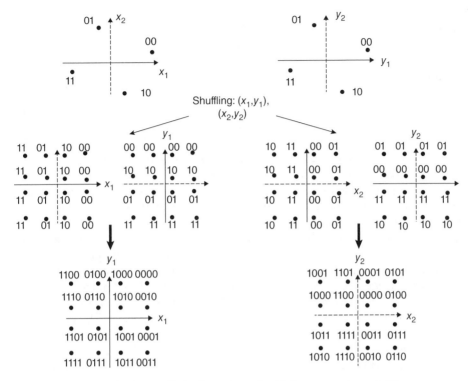

Figure 6.7 16-point SCMA constellation for codewords with 2 non-zero elements [8].

part of the points of a complex mother constellation. Layer specific operators are then applied to the mother constellation to build codebooks for layers of SCMA. Simulation results illustrate the gain of SCMA over LDS and OFDMA for fading channels. SCMA demonstrates all the multi-access benefits of LDS in terms of overloading, moderate detection complexity, interference whitening. Potentially SCMA will achieve good system performance in both uplink and downlink multiple access scenarios in the future wireless networks.

In [13] and [14], an optimized construction of SCMA codebooks is proposed for downlink SCMA system. The work aims to find new mother codebook based on star-QAM signalling constellation. The SCMA users' codebooks are then generated from the mother codebook by using some necessary operators. The results indicate that the new codebooks can outperform LDS and the existing codebook presented in [8]. The authors in [15] investigate the application of spherical codes in an SCMA system, where the spherical codes are used to build multi-dimensional mother constellations for SCMA codebooks. Spherical codes are easy to perform and have large coding gain, so that it is suitable for use on SCMA systems. The simulation results show that the proposed SCMA codebooks have performance advantages over known SCMA codebooks in [8]. The work in [16] proposes the multi-user codebook design for SCMA system over Rayleigh fading channel. A novel and efficient performance criterion is proposed by rotated constellation design to optimize the mutual information and the cut-off rate of equivalent multiple-inputs multiple-outputs (MIMO) system. The simulation results presented that the proposed scheme provides a substantial block error rate compared to [8] and [15].

The paper [17] mentioned another way to design the SCMA codebook based on the constellation rotation and interleaving method for downlink SCMA systems. The main idea is to define the first dimension of mother constellation by a subset of lattice $\mathbf{Z}^2$; then the other dimensions can be obtained by rotating the first dimension. Then a one-to-one Gray mapping from the constellation point to the codeword of multi-dimensional mother codebook is defined, and the elements of even dimension are interleaved to improve the performance in fading channel. The results show that the Bit Error Rate (BER) performance of multi-dimensional SCMA codebooks outperforms the SCMA codebooks in [8] and LDS [18] in downlink Rayleigh fading channels.

The researchers in [19] jointly consider codebook design (mapping matrix and constellation graph design) and codebook assignment to investigate the detection complexity minimization problem for uplink SCMA networks. A cost-efficient algorithm based on the dual coordinate search method (DCSA) is proposed to solve the integer program in a more direct way to improve the efficiency. The paper also designs the constellations for each codebook to reduce detection complexity. An example of the mother constellation design is given in Figure 6.8 (a) and (c) shows how the traditional one-dimensional modulation represents two- and three-bit information while (b) and (d) demonstrate that the complexity can be reduced from $O(4^d)$ to $O(3^d)$ and from $O(8^d)$ to $O(5^d)$ respectively. Based on such a mother constellation, the constellation for each codebook can be constructed through complex conjugate, phase rotation or dimensional permutation. The work demonstrates huge complexity reduction for detection in comparison with other benchmarking schemes.

In order to improve the multi-user spectrum efficiency for SCMA systems, the authors in [20] proposed an offline codebook design method from the perspective of multi-user information theory by maximizing the constellation constrained capacity. The

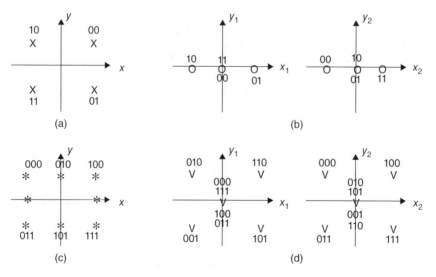

Figure 6.8 Mother constellation design for representing two or three bits in traditional and SCMA systems [19].

multi-dimensional codebook is constructed by optimizing a series of one-dimensional constellation sets. Constellation rotation between different users is utilized to achieve better superposition constellation constrained capacity with the help of the modified multi-start interior-point algorithm. Through link-level simulations, the results show that the proposed codebook overperforms the previously known schemes, SCMA codebook [8], the star-QAM codebook [13] and ultra- dense LDS [21] by 0.5dB, 0.5dB and 1dB respectively in terms of constellation constrained capacity, and 2.2dB, 1.8dB and 3dB gain respectively in terms of BER.

6.3.2 Decoding/Detecting Techniques for SCMA

The SCMA decoder is a crucial part of the whole decoding process. The message passing algorithm (MPA) is the first decoding method that is used in SCMA, which is first proposed in [7] for CDMA. This algorithm provides good performance on SCMA decoding, but its complexity is experiencing exponential growth with the number of users on one physical resource element (PRE). In [22], the authors proposed the logarithm domain MPA (Log-MPA) to reduce decoding complexity. There, most multiplication and exponent operations have been turned into simple addition and maximization, which can save computational time and facilitate hardware implementation. The simulation and experimental results show the comparison of the complexity of Log-MPA and MPA refers to the number of operations and running time in the prototype system. It illustrates that the Log-MPA saves more than 90% multiplication and eliminates the exponent calculations. Moreover, it has been shown to save more than 2/3 running time compared with MPA scheme.

In [23], the authors asserted that the improvement of Log-MPA is limited due to the calculation of Euclidean distance and iterations still needing to be calculated. Thus they proposed the weighted message passing algorithm (WMPA) for SCMA decoding to reduce computation complexity by allocating larger weight factors to those codewords with larger probability values, and smaller weight factors to smaller ones. The results show that it can provide a better block error rate when the signal-to-noise-ratio (SNR) and iteration number are larger. This logarithm also significantly reduced the decoding time compared with the Log-MPA and MPA methods.

The authors in [24] proposed a fixed low complexity MPA scheme based on partial marginalization (PM-MPA), this scheme can enable quite an acceptable uplink multi-user detection with massive connectivity. The results illustrated that the implementation complexity of this scheme can reduce the computational complexity compared with the original MPA decoder, but it also suffers from BER performance losses. In [25], the author presented a shuffled MPA (S-MPA) scheme based on message updated strategy. This scheme can greatly reduce complexity by accelerating the convergence rate and in the meantime it can provide a lower BER performance compared with MPA and PM-MPA.

In addition, the authors in [26] proposed a novel MPA detection scheme based on sphere decoding called SD-MPA to reduce the decoding complexity; due to the existence of AWGN, the received superposed constellation point will not overlap with transmitted superposed constellation point in constellation domain. By just calculating these points inside a circular region with defined radius according to AWGN variance, the proposed SD-MPA can dynamically make a trade-off between performance and complexity.

In order to reduce the decoder computational complexity of multiple-input multiple-output (MIMO) SCMA systems, the authors in [27] proposed a decoding algorithm – Partial Decoding MPA (PD-MPA) – for a space frequency block code (SFBC) based MIMO SCMA system. This algorithm can dynamically remove the number of combinations in the update at resource mode. The simulation results show that PD-MPA algorithm has similar BLER performance to MPA and lower computation complexity.

[28] proposed an optimal receiver for SCMA based on the sphere decoding (SD); the corresponding decoding scheme is called the list sphere decoding (LSD) method. The authors introduced the SD method into the SCMA system to realize the optimal multi-user detection by employing the Schnorr–Euchner (SE) enumeration strategy [29]. The implementation of computation complexity is much lower than MPA. The simulation results regarding system performance and computation complexity analysis are evaluated on AWGN and the fading channel to see that the proposed optimal receiver provides notable gain compared with the MPA.

The authors in [30] proposed an adaptive decoder based on deterministic MPA(DMPA), algorithmic simplifications such as domain changing and probability approximation. In addition, early termination and initial noise reduction are also employed for faster convergence and better performance. This work also is implemented on Xilinx Virtex-7 XC7VX690T FPGA to demonstrate its advantages for real applications. The numerical results show that the proposed optimization benefit both decoding complexity and speed. Compared with other methods, it has the lowest computational complexity while maintaining the error performance.

Table 6.1 provides the comparison of above schemes with regard to the number of operations and the BER performance.

6.3.3 Other Research on Performance Evaluation of SCMA

The authors in [9] evaluate the performance of SCMA via system-level evaluations for a small packet application scenario. In LTE, uplink transmission is scheduled in the base station (BS) with a request-grant procedure. A UE sends a scheduling request (SR) to the network on uplink dedicated resources that occur periodically, normally every 5 or 10 ms. Even in the best case, if the data arrive at the UE right before the SR, it needs another 7 ms between the request and transmission [31]. This is not good for delay-sensitive and busy traffic of future applications. The evaluation focuses on the uplink connections for small packet transmission with low latency of less than 5 ms. To evaluate the performance of data transmissions for critical information (e.g., short-range warning messages), a packet is considered as "lost" if its transmission failed at the first attempt. The simulation results demonstrate that SCMA outperforms OFDMA in terms of supported number of active users when achieving the same system outage. And clearly, reveal that the packet drop rate increase as more packets are to

Table 6.1 Comparisons of different decoding schemes with regard to the number of operations and BER performance.

Number of operations	MPA > PM-MPA > Log-MPA > S-MPA > SD-MPA
BER	Log-MPA > MPA > WMPA > PM-MPA > LDS> S-MPA > SD-MPA

be delivered in both SCMA and OFDMA. However, the OFDMA packet drop rates are more sensitive to the change of load. In addition, SCMA demonstrates much lower average packet drop rate than that in the OFDMA with the same packet arrival rate. The simulation results demonstrate that SCMA outperforms OFDMA in terms of supported number of active users when achieving the same system outage. Benefits of utilizing SCMA in large-scale heavy loaded scenarios are demonstrated in [9]: SCMA not only naturally supports a large number of users, but is also capable of maintaining good packet dissemination.

The authors in [32] proposed a resource allocation scheme to increase the number of successful accesses by applying the access class barring (ACB) scheme and SCMA technique for massive machine-type communication devices (MTCDs). This scheme employs ACB to control the number of random access (RA) attempts during the preamble transmission in physical random access channel (PRACH), then allocates SCMA codebook resource in the physical uplink shared channel (PUSCH) to those MTCDs with successful preamble transmission. A joint PRACH and PUSCH resource allocation optimization problem is also formulated. Due to the features of MTC systems, small-sized (e.g., 80 bits per packet) data transmissions are considered in the simulations, which resembles the traffic flows for safety-related use cases in vehicular networks. The paper demonstrates that the proposed scheme can achieve the largest successful accesses compared to other benchmarking random access schemes for MTC. The results illustrate that the scheme combining the proposed access class barring (ACB) and SCMA technique outperforms other benchmarking schemes. The gain increases as the SNR increases.

The authors in [11] investigate the system capacity in terms of link-level performance. The Block Error Rate (BLER) vs SNR result clearly shows that the system is overloaded by 150% when 6 UEs are using 6 layers ($J=6$) and 4-point codebooks ($N=4$) separately. In the 300% overloading case, 12 UEs share 6 layers, where 2 UEs use the same layer (codebook) with the different pilot sequence. It can be seen that the overloading capacity may ensure the massive connectivity (many more UEs) utilizing the same limited radio resource. The system capacity of NB-SCMA can be as high as more than 300% of the existing non-overlapping based systems without link-level performance degradation.

The authors in [33] present a codebook-based resource allocation algorithm for uplink networking using the matching game. The BS assigns to each user a set of sub-carriers corresponding to a specific codebook, and each user performs power control over multiple sub-carriers. This work aims to optimize the sub-carrier assignment and power allocation to maximize the total sum-rate. To deal with this problem, a novel swap-matching algorithm is proposed. The performance of proposed SCMA and PFDMA are evaluated, and the authors show that the SCMA scheme outperforms the traditional OFDMA scheme in terms of total sum-rate and the overloading gain.

To meet the low-latency requirements in the 5G mobile network, the tedious procedure of link establishment and scheduling could be removed from grant-free random access to shorten overall latency. SCMA is a promising scheme for consideration in random access since its non-orthogonality enables the BS to accommodate many more users compared with other multiple access schemes. Retransmission is used for achieving the desired reliability. However, too many retransmission times may undercut the advantage and bring in long latency. In [34], the authors proposed an improved hybrid automatic repeat request (HARQ) scheme in SCMA under grant-free random access scenario. Firstly, the retransmitting users randomly pick a back-off time in the given delay window to leverage the pile-up effect of users' data. Secondly, a codebook-hopping scheme introduced in [35] is used to resolve the potential codebook collision problem in random access, so the package error rate can be reduced. Simulation results show that the SCMA with the proposed HARQ scheme increases the throughput and reduces the latency by 20.21% and 24.69% compared with the SCMA with traditional HARQ scheme.

In [36], the authors proposed an ACK feedback-based UE to CTU mapping rule for uplink grant-free transmission, where the UE who has retransmission data can get an exclusive radio resource to ensure an successful retransmission. The simulation results indicate that the proposed scheme can decrease the collision probability and reduce the number of retransmissions to save energy.

The authors in [37] proposed a downlink resource allocation strategy in SCMA system based on two widely used algorithms: Proportional Fair (PF) algorithm and Modified Largest Weighted Delay First (M-LWDF) algorithm. The results demonstrate that the SCMA-based strategy performs better compared with OFDM-based strategy in terms of communication capacity, throughput, time delay, packet loss rate and fairness.

The work in [38] presents a resource allocation algorithm for V2X communications based on SCMA to maximize the system throughput. By analysing the interference model in the V2X scenario, the authors combined a graph colour-based user cluster algorithm and resource allocation algorithm to solve the problem. In the simulation, the authors considered a hybrid network consisting of cellular user and vehicle user pairs in a single cell and adopt SCMA for cellular users. The throughput performance of different access methods (SCMA and OFDMA) shows that SCMA has a better performance than OFDMA. Due to the overload feature, SCMA allows more users to access the system, and leads to a larger average throughput.

One strong assumption in literature is that all the codebooks are predefined by a centralized controller and are available for access by all the users. This assumption may be not practical in scenarios where a new user joins and leaves the system fast and frequently due to high mobility. This affects the system performance in networks for fast-moving vehicles. In addition, signalling overhead and latency issues in the realization of such provision should be studied in future vehicular communication systems.

Table 6.2 summarizes the highlights and findings in the state-of-the-art efforts on SCMA evolvement. In addition, it also provides the links to CAV to show the advantages and research gaps for the future work.

6.4 Conclusion and Future Work

Considering the dramatic growth of traffic, 5G communication technology is proposed to support massive connectivity with a large number of devices. CAVs network is one of the important future applications that can benefit from 5G. SCMA, as a promising technique in 5G wireless communication networks, has recently received lots of attention. In this chapter, an overview of the current design of SCMA systems, and the most representative works in terms of system enhancement are presented. Table 6.2 summarizes the highlights and findings in the state-of-the-art efforts on SCMA evolvement. In this table, the suitability of SCMA-based approaches in CAV systems is also illustrated,

Table 6.2 Summary of current representative efforts on SCMA systems.

Focus	Reference	Highlights	Findings	Links to CAVs
Codebook redesign	[4]	1. Unified framework; 2. Multi-stage sub-optimal systematic approach; 3. Evaluation of overloading capability and reliability.	SCMA is proposed for the first time to show the low complexity reception technique but better performance compared with LDS; SCMA supports multiple times more users than OFDMA.	Massive devices; small packets; Reliability is improved.
	[8]	1. Lattice constellation; 2. Constellation with low projections; 3. Link-level performance evaluation compared with LDS and OFDMA.	Unitary rotation control, dimensional dependency and power variation while keeping the Euclidean distance unchanged; SCMA BLER outperforms LDS, OFDMA and single carrier FDMA and the gain is over 2dB compared to OFDMA and single carrier FDMA.	
	[13], [14]	1. Star-QAM based multi-dimensional signalling constellation; 2. BER performance evaluated and compared with LDS and scheme in [8].	New mother codebook has been investigated, the power of each codeword in each codebook is different; The new codebook design can greatly improve the BER performance without sacrificing the low detection complexity, compared with LDS and [8].	
	[15]	1. Spherical codes used to build multi-dimensional mother constellation; 2. Packet error rate is evaluated and compared with the results in [8].	The spherical code is considered in SCMA codebook design; The results show that the spherical codes C(3,4) and C(4,16) have coding gains 1.25dB and 0.88dB over known SCMA codebooks T4QAM [8] and T16QAM [39], respectively.	

(Continued)

Table 6.2 (Continued)

Focus	Reference	Highlights	Findings	Links to CAVs
	[16]	1. A rotated multi-dimensional constellation proposed based on optimizing the mutual information and cut-off rate; 2. BLER performance is investigated and compared with the schemes in [8] and [15].	The proposed criterion is efficient in developing codebooks; The BLER performance of the proposed scheme has a gain about 0.8dB over [8] and a gain about 0.5 dB over [15].	
	[17]	1. Codebook design based on constellation rotation and interleaving method for downlink SCMA systems; 2. BER performance of proposed codebook compared with LDS and [8].	Codebook designed as spectral or power efficiency; BER performance shows that the proposed codebook design outperforms the LDS and the work in [8], where the 0.6dB improvement can be achieved.	
	[19]	1. Redesign the mother constellation and codebook assignment algorithm; 2. simulation results with regard to detection complexity compared with other schemes.	Codebook assignment algorithm DCSA helps improve spectrum efficiency; Mother constellation design reduces detection complexity by 50 times and more.	Reduced detection complexity reduces processing latency which is needed for future CAVs networking.
	[20]	1. Multi-step codebook design method for maximizing the constellation constrained capacity is proposed; 2. BER performance produced and compared with LDS and the work in [8], [13], [21].	The BER performance of the proposed design already approaches to Shannon limit when the SNR is lower than 3dB and achieves significant gains of 2.2dB, 1.8dB and 3dB compared with [8], [13], [21] respectively.	Performance approaches to the Shannon limit which is suitable for use in the CAV networks.
Decoding/ Detection methods	[22]–[28] [30]	1. Different decoding or detection methods are proposed; 2. Comparisons are given from operations numbers of running time or BER performance.	Performance comparison results are given in Table 6.1.	These studies are dedicated to reducing the decoding and detection complexity and thus reduce the latency, which is useful for the future design of 5G communication networks for CAVs.

Topic	Ref.	Details	Results	Remarks
Blind detection	[10]	1. Blind detection technique designed for SCMA-based uplink grant-free multiple access scenarios; 2. Performance in terms of detection missing rate and BLER are evaluated.	Promising technology to enable massive connectivity for grant-free multiple access transmissions. The simulation results show that the proposed expectation maximization can achieve a misdetection probability as low as 0.5%.	Grant-free multiple access algorithm reduces signalling overhead and latency; supportive to massive connections.
Uplink SCMA contention-based access	[9]	1. Contention-based SCMA design is proposed to support massive connectivity; 2. Capacity and QoS evaluation for SCMA-based uplink connections.	Achieving about 300% capacity over OFDMA in the reference scenarios; SCMA is more reliable than OFDMA in terms of packet drop rate. The results clearly show that the OFDMA packet drop rate degrades much faster with the increase of loads than SCMA.	Improved reliability; massive connectivity.
Narrow-band SCMA scheme	[11]	1. Narrow-band SCMA scheme is proposed, where overloading capability is measured for users on the same layer using different pilot sequences; 2. BLER performance is evaluated and compared with narrow-band LTE.	Overloading capacity ensures massive connectivity without link-level performance degradation utilizing the same limited radio resource.	Supportive to massive connections.
Random access	[32]	1. Uplink transmissions for MTC devices; Access Class Barring to control the number of massive RA attempts during preamble transmission in PRACH; 2. Optimize the length of SCMA codeword to maximize the number of successful accesses.	Improved channel access in terms of number of successful accesses under different SNRs.	Small-sized data (80 bits per packet) resembles traffic flows in safety-related use cases; improved reliability.
Resource allocation	[33]	1. Codebook-based resource allocation scheme is proposed based on the swap-matching algorithm; 2. The performance in terms of total sum-rate is evaluated and compared with OFDMA and random allocation scheme.	The proposed algorithm outperforms the traditional OFDMA scheme in terms of overloading gain.	Improving the V2X networking access rate and balancing the detecting complexity and sum rate.

(Continued)

Table 6.2 (Continued)

Focus	Reference	Highlights	Findings	Links to CAVs
Random access	[34]	1. Improved HARQ scheme in SCMA under grant-free random access scenario; 2. Simulation results with regard to retransmission count, throughput and PER are presented.	The new scheme considered the balance between the retransmission latency and information reliability; The results indicate that compared with the DCMA with traditional HARQ scheme, the proposed scheme increases the throughput by 20.21% and reduces the average retransmission latency by 24.69%.	Considering 5G in future CAVs communication networks, the benefit of this scheme is reducing the latency, but it also can maintain the reliability which is important for future CAVs.
ACK in resource allocation	[36]	1. ACK feedback-based UE to CTU mapping rule for uplink grant-free transmission, so the UE who has retransmission data can get an exclusive radio resource to ensure a successful retransmission; 2. Collision possibility is evaluated and compared with conventional schemes.	A new UE to contention transmission unit mapping rule is proposed which applies ACK to indicate the radio resource allocation; Compared with conventional mapping rules, the collision probability is reduced by 20% after applying the new rule.	The new scheme can decrease the collision probability and reduce the number of retransmissions. It also can save the energy, which is suitable for future 5G adapted CAVs networks.
Resource allocation	[37]	1. SCMA-based downlink resource allocation strategy is designed based on PF and M-LWDF; 2. Throughput, packet loss rate and latency are compared with OFDMA-based resource allocation scheme.	The system throughput has been improved and the packet loss rate and latency have been reduced. Under proportional fair algorithm, the SCMA platform can achieve 3Mbps more system throughput compared with the OFDMA platform when the number of users reaches 100. In addition, the packet loss rate and average delay are reduced by 43% and 40% respectively compared with OFDMA platform.	Latency reduction which is a benefit for future massive connectivity V2X communications with CAVs.
Resource allocation	[38]	1. Resource allocation scheme for V2X based on SCMA is proposed by combining a graph colour-based user cluster and resource allocation algorithm; 2. Throughput performance is evaluated and compared with OFDMA.	A hybrid network consisting of cell users and vehicle users in a single cell is considered; The throughput of the SCMA-based system is increased by 40% compared with the OFDMA-based system.	An SCMA-based resource allocation scheme is applied to V2X communications shows better performance than OFDMA.

which shows the advantages and research gaps for the future work. The outcomes of the studies of the existing efforts on SCMA-based approaches can be summarized as follows: firstly, SCMA will support massive users and frequent small-sized packet transmissions. Secondly, it improves reliability compared with traditional techniques such as OFDMA and LDS, but still does not fulfil the requirements of CAV use cases. To overcome this problem, error detection and correction can be applied at the receiver side to improve the reliability of data transmission in CAV use cases (e.g., automated overtake, high density platooning). Finally, the latency (such as processing delay and delay due to contention) could be reduced by reducing the detection complexity and/or overloading capacity.

To conclude, the characteristics of the SCMA technique show that it is applicable in future intelligent transportation systems. Further potential work may be on the use of SCMA in resource allocation schemes to improve the access rate in future intelligent transportation systems with CAVs.

References

1 Department for Transport (ed.) Safety of loads on vehicles: code of practice. Department for Transport, London, 2002.

2 H. Wei. 5G: new air interface and radio access virtualization. White Paper, 2015.

3 F. N. Julio A. Sanguesa, Vicente Torres-Sanz, Manuel Fogue, Piedad Garrido, and Francisco J. Martinez. On the study of vehicle density in intelligent transportation systems. *Mobile Information Systems*, pages 1–13, 2016.

4 H. Nikopour and H. Baligh. Sparse code multiple access. In *2013 IEEE 24th Annual International Symposium on Personal, Indoor, and Mobile Radio Communications (PIMRC)*, 2013, pages 332–336.

5 Y. Saito, Y. Kishiyama, A. Benjebbour, T. Nakamura, A. Li, and K. Higuchi. Non-Orthogonal Multiple Access (NOMA) for cellular future radio access. In *2013 IEEE 77th Vehicular Technology Conference (VTC Spring)*, 2013, pages 1–5.

6 G. Miao, J. Zander, K. W. Sung, and S. B. Slimane, *Fundamentals of Mobile Data Networks*. Cambridge University Press, 2016.

7 R. Hoshyar, F. P. Wathan, and R. Tafazolli. Novel low-density signature for synchronous CDMA systems over AWGN channel. *IEEE Transactions on Signal Processing*, vol. 56, pages 1616–1626, 2008.

8 M. Taherzadeh, H. Nikopour, A. Bayesteh, and H. Baligh. SCMA Codebook Design. In *2014 IEEE 80th Vehicular Technology Conference (VTC2014-Fall)*, 2014, pages 1–5.

9 K. Au, L. Zhang, H. Nikopour, E. Yi, A. Bayesteh, U. Vilaipornsawai, et al. Uplink contention based SCMA for 5G radio access. In *2014 IEEE Globecom Workshops (GC Wkshps)*, 2014, pages 900–905.

10 A. Bayesteh, E. Yi, H. Nikopour, and H. Baligh. Blind detection of SCMA for uplink grant-free multiple-access. In *Wireless Communications Systems (ISWCS), 2014 11th International Symposium on*, 2014, pages 853–857.

11 J. Wang, C. Zhang, R. Li, G. Wang, and J. Wang. Narrow-band SCMA: a new solution for 5G IoT uplink communications. In *Vehicular Technology Conference (VTC-Fall), 2016 IEEE 84th*, 2016, pages 1–5.

12 M. Alam and Q. Zhang. Performance study of SCMA codebook design. In *2017 IEEE Wireless Communications and Networking Conference (WCNC)*, 2017, pages 1–5.

13 L. Yu, X. Lei, P. Fan, and D. Chen. An optimized design of SCMA codebook based on star-QAM signaling constellations. In *2015 International Conference on Wireless Communications & Signal Processing (WCSP)*, 2015, pages 1–5.

14 L. Yu, P. Fan, X. Lei, and P. T. Mathiopoulos. BER analysis of SCMA systems with codebooks based on Star-QAM signaling constellations. *IEEE Communications Letters*, vol. 21, pages 1925–1928, 2017.

15 J. Bao, Z. Ma, M. A. Mahamadu, Z. Zhu, and D. Chen. Spherical codes for SCMA codebook. In *2016 IEEE 83rd Vehicular Technology Conference (VTC Spring)*, 2016, pages 1–5.

16 J. Bao, Z. Ma, Z. Ding, G. K. Karagiannidis, and Z. Zhu. On the design of multiuser codebooks for uplink SCMA systems. *IEEE Communications Letters*, vol. 20, pages 1920–1923, 2016.

17 D. Cai, P. Fan, X. Lei, Y. Liu, and D. Chen. Multi-dimensional SCMA codebook design based on constellation rotation and interleaving. In *Vehicular Technology Conference (VTC Spring), 2016 IEEE 83rd*, 2016, pages 1–5.

18 J. v. d. Beek and B. M. Popović. Multiple access with low-density signatures. In *GLOBECOM 2009– 2009 IEEE Global Telecommunications Conference*, 2009, pages 1–6.

19 D. Zhai, M. Sheng, X. Wang, and J. Li. Joint codebook design and assignment for detection complexity minimization in uplink SCMA networks. In *2016 IEEE 84th Vehicular Technology Conference (VTC-Fall)*, 2016, pages 1–5.

20 K. Xiao, B. Xia, Z. Chen, B. Xiao, D. Chen, and S. Ma. On capacity-based codebook design and advanced decoding for sparse code multiple access systems. *IEEE Transactions on Wireless Communications*, pages 1–1, 2018.

21 A. R. Safavi, A. G. Perotti, and B. M. Popović. Ultra low density spread transmission. *IEEE Communications Letters*, vol. 20, pages 1373–1376, 2016.

22 S. Zhang, X. Xu, L. Lu, Y. Wu, G. He, and Y. Chen. Sparse code multiple access: An energy efficient uplink approach for 5G wireless systems. In *2014 IEEE Global Communications Conference*, 2014, pages 4782–4787.

23 D. Wei, Y. Han, S. Zhang, and L. Liu. Weighted message passing algorithm for SCMA. In *2015 International Conference on Wireless Communications & Signal Processing (WCSP)*, 2015, pages 1–5.

24 H. Mu, Z. Ma, M. Alhaji, P. Fan, and D. Chen. A fixed low complexity message pass algorithm detector for up-link SCMA system. *IEEE Wireless Communications Letters*, vol. 4, pages 585–588, 2015.

25 Y. Du, B. Dong, Z. Chen, J. Fang, and L. Yang. Shuffled multiuser detection schemes for uplink sparse code multiple access systems. *IEEE Communications Letters*, vol. 20, pages 1231–1234, 2016.

26 L. Yang, X. Ma, and Y. Siu. Low complexity MPA detector based on sphere decoding for SCMA. *IEEE Communications Letters*, vol. 21, pages 1855–1858, 2017.

27 Z. Wu, C. Zhang, X. Shen, and H. Jiao. Low complexity uplink SFBC-based MIMO-SCMA joint decoding algorithm. In *2017 3rd IEEE International Conference on Computer and Communications (ICCC)*, 2017, pages 968–972.

28 G. Chen, J. Dai, K. Niu, and C. Dong. Optimal receiver design for SCMA system. In *Personal, Indoor, and Mobile Radio Communications (PIMRC), 2017 IEEE 28th Annual International Symposium on*, 2017, pages 1–6.

29 C.-P. Schnorr and M. Euchner. Lattice basis reduction: Improved practical algorithms and solving subset sum problems. *Mathematical programming*, vol. 66, pages 181–199, 1994.

30 C. Zhang, C. Yang, W. Xu, S. Zhang, Z. Zhang, and X. You. Efficient sparse code multiple access decoder based on deterministic message passing algorithm. *arXiv preprint arXiv:1804.00180*, 2018.

31 E. Dahlman, S. Parkvall, and J. Skold, *4G: LTE/LTE-advanced for mobile broadband*: Academic Press, 2013.

32 T. Xue, L. Qiu, and X. Li. Resource allocation for massive M2M communications in SCMA network. In *Vehicular Technology Conference (VTC-Fall), 2016 IEEE 84th*, 2016, pages 1–5.

33 B. Di, L. Song, and Y. Li. Radio resource allocation for uplink sparse code multiple access (SCMA) networks using matching game. In *2016 IEEE International Conference on Communications (ICC)*, 2016, pages 1–6.

34 J. Lian, S. Zhou, X. Zhang, and Y. Wang. An improved HARQ scheme for SCMA under random access. In *2017 3rd IEEE International Conference on Computer and Communications (ICCC)*, 2017, pages 1163–1167.

35 J. LIAN, S. ZHOU, X. ZHANG, and Y. WANG. An SCMA control channel accessing scheme based on codebook-hopping under hyper-cellular network architecture. *SCIENTIA SINICA Informationis*, vol. 47, pages 789–799, 2017.

36 J. Shen, W. Chen, F. Wei, and Y. Wu. ACK feedback based UE-to-CTU mapping rule for SCMA uplink grant-free transmission. In *Wireless Communications and Signal Processing* (WCSP), 2017 9th International Conference on, 2017, pages 1–6.

37 W. Liqiang, L. Xinyue, G. Shan, and H. Xingling. Resource allocation strategy in SCMA based multiple access system. In *Electronics Instrumentation & Information Systems* (EIIS), 2017 First International Conference on, 2017, pages 1–6.

38 W. Wu and L. Kong. Resource allocation algorithm for V2X communications based on SCMA. *arXiv preprint arXiv:1804.03529*, 2018.

39 M. T. Boroujeni, H. Nikopour, A. Bayesteh, and M. Baligh. System and method for designing and using multidimensional constellations. Google Patents, 2016.

7

5G Communication Systems and Connected Healthcare

David Soldani and Matteo Innocenti

This chapter focuses on 5G mobile systems for ultra-reliable low latency communications applied to two specific healthcare use cases: Wireless Tele Surgery (WTS), using a mobile console and a robotic platform with video, audio and haptic feedback; and Wireless Service Robot (WSR), a companion robot or service robot performing tasks of social caretakers, professional personnel or family members. Cyber-sickness and latency between the robotic platform and a reasoning server are the new technical problems the 5G mobile system is expected to solve, e.g. in hospices, hospitals, homes and campus areas.

In the following sections, we first present the use cases and related technical requirements; then, we describe the 5G reference architecture and key enabling technologies to meet the performance, dependability and security targets for Tele-healthcare. This is followed by a comprehensive analysis of the costs and revenues operators and healthcare providers would encounter to deliver the corresponding services, if a 5G system was deployed.

7.1 Introduction

The next generation of mobile broadband infrastructure or International Mobile Telecommunications (IMT) for 2020 and beyond (5G) will expand and support diverse usage scenarios and applications with respect to previous network generations, purposed primarily for the support of improved voice, mobile Internet and video experience. The main categories of usage scenarios for 5G are [1–3]: *enhanced Mobile Broadband* (eMBB), addressing human-centric use cases for access to multimedia content, services and data; *Ultra-Reliable Low-Latency Communications* (URLLC), with strict requirements, especially in terms of end-to-end latency and reliability; and *massive Machine-Type Communications* (mMTC), for a huge number of connected devices, typically transmitting a relatively low volume of non-delay-sensitive information.

Enabling 5G Communication Systems to Support Vertical Industries, First Edition.
Edited by Muhammad Ali Imran, Yusuf Abdulrahman Sambo and Qammer H. Abbasi.

In this chapter, we focus on URLLC for connecting remote-controlled robots or robotic platforms with intelligence in the edge cloud, i.e. a point of presence (PoP), namely a data centre, close to the robots [4].

The first use case is *Wireless Tele Surgery*, which is considered to be an integral part of the wider field of telemedicine. The aim of telemedicine is to provide specialized healthcare services over long distances to improve the quality of life of patients located in isolated areas, where access to specialized medical services is limited. With 5G, a specialist could examine or operate on a patient at a different geographic location, and the system could potentially remove barriers to healthcare provision in developing countries, in areas of natural disasters, and in war zones, where consistent healthcare services are unavailable or there is no time to transport a patient to a hospital. WTS is primarily about the case where a surgeon in one remote location performs an operation on a patient laid in another room with the aid of a robotic platform [5]. The first successful tele surgery, named "Operation Lindbergh", was performed using a Zeus robotic system in 2001 [6]. Since then, tele surgery has been witnessing tremendous growth [7]. Basically, it consists of a short- or long-distance system that utilizes wired and/or wireless communication networks. A representative paradigm of the short-distance case is the Da Vinci Surgical System. The concept of long-distance tele-robotics is typically exemplified by diagnostic ultrasound scanning systems. In WTS, the aim of introducing 5G technology is to develop Teleoperated surgical systems, which actually was not feasible with the deployment of previous wireless technologies [8]. The target with 5G is to keep latency – the time it takes to successfully deliver an application layer packet or message between wireless communication-rich hospital campuses – below certain thresholds and use robotic surgical devices, which can be operated by an expert surgeon based hundreds miles away from the patient. With the deployment of 5G wireless, the medical doctor may perform remote robotic surgery, or other tasks, e.g. medical ultrasound in a decentralized hospital system at any hospital within such a system; also, using 5G, he or she can be consulted at any time from anywhere. In general, robotic surgery reduces hospitalization time, pain and discomfort and recovery time; it infers smaller incisions, reduced risk of infection, reduced blood loss and transfusions, minimal scarring and better quality, e.g. see [9] and [10]. In order to reach the same "sense of touch" as in conventional interventions, without the remote doctor experiencing *cyber-sickness*, the 5G system must provide the guaranteed low latency and the necessary stability while transmitting *haptic feedback* (tactile and/or kinaesthetic signals), and improved throughput for better visualization, enhanced dexterity, and greater precision, well beyond today's Da Vinci system.

The second use case is about connecting a *Wireless Service Robot*, i.e. a robotic platform with Artificial Intelligence (AI) in the edge cloud [11]. The target is to stay below and above the required latency and throughput, respectively, with a very high degree of reliability between the wireless robot and edge computing servers, and use robotic care devices to assist patients and old people in hospitals, hospices and in their campus areas, and, mostly, at home. The edge cloud (reasoning system) runs computer vision algorithms to interpret human emotions, enable the service robot to interact naturally and perform complex care or household tasks. The main objective is to reduce care costs and, especially, help people to *remain active and independent with a good quality of life*. The 5G wireless connectivity will replace cables, reduce costs, and thus enable a massive adoption and utilization of robotic platforms, globally.

7.2   Use Cases and Technical Requirements

In the following sections, we describe the WTS and WSR use cases in detail and present the corresponding technical requirements 5G wireless is expected to support.

7.2.1   Wireless Tele Surgery

As shown in Figure 7.1, the target is to provide a remote surgeon, who could be located hundreds of kilometres away from patients, with the same sense of touch (essential for localizing hard tissue or nodules) while substituting doctor's hands in interventions with robotic probes (arm or finger). In order to achieve such an experience in remote interventions, delay and stability are the crucial parameters in transmitting haptic feedback (kinaesthetic info, such as force or motion; and/or tactile indicators, such as vibration or heat), in addition to audio/video data, because delays can seriously impair the stability of the feedback process and lead to *cyber-sickness*, which may occur when our eyes observe a movement which is delayed compared to what our vestibular system perceives [12–15].

A reaction time of about 100 ms, 10 ms and 1 ms is required for auditory, visual and manual interaction, respectively [14]. In order to realize these reaction times, so that all human senses can, in principle, interact with machines, and hear and see things far away, the end-to-end Quality of Service (QoS) needs to be guaranteed in terms of reliability (failure rate even below 10^{-7}, i.e. 3.17s of outage per year), speed (30–50 Mb/s encoded bit rate/view, or 1 Gb/s encoded bitrate for holographic rendering) and latency (below 100 ms, 25 ms and 5 ms for audio/video feedback, haptic feedback and interactive live holographic feedback, respectively) [12–15], as depicted in Figure 7.2. The 5G system requires a limited distance between sites, user plane (UP) in local break-out on the edge and a dedicated slice, which isolates traffic and ensures Service Level Agreement (SLA). (Both public and enterprise network options were presented in [4].)

7.2.2   Wireless Service Robots

Service robots for care are being primed to join the labour force in various roles, e.g. logistic, cleaning and monitoring, which can be fully automated. Beyond these simple

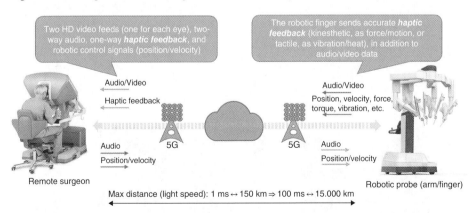

Figure 7.1 WTS: Examples of signals and parameters exchanged between a remote surgeon and a local robotic probe.

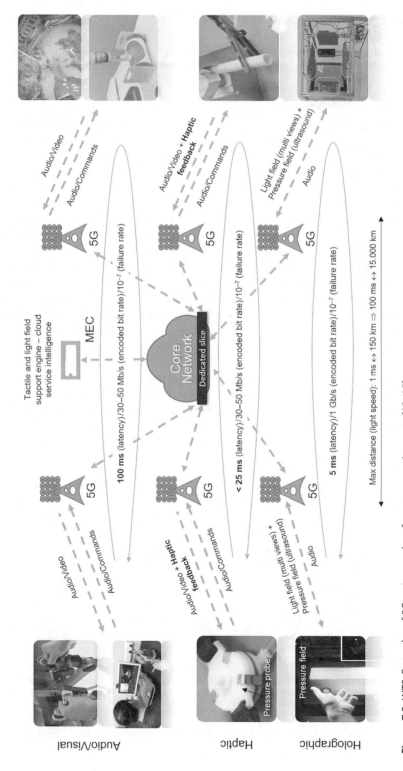

Figure 7.2 WTS: Examples of 5G system and performance requirements [12–15].

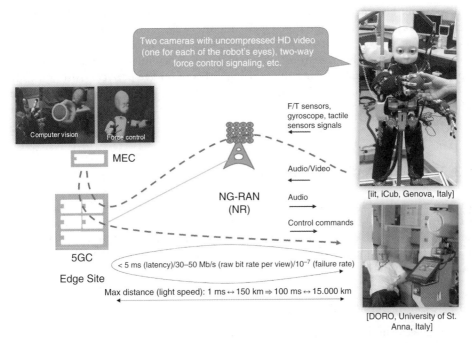

Figure 7.3 WSR: Examples of signals exchanged between the MEC server and robot and performance requirements [19, 20].

tasks, androgynous robots are anticipated to interpret human emotions, to interact naturally with people and perform complex care or household jobs, as well as to assist patients and old people in hospital and hospice campus areas, and at home to reduce care costs, and to help ageing people to remain active and independent with good quality of life as long as possible [16–20].

The target with 5G wireless is to meet the required latency and throughput with ultra-high reliability, between a wireless robot and a *Multi-access Edge Computing* (MEC) server (reasoning system) [21], where most of the intelligence is located, e.g. for object tracking, recognition and related applications [19, 20], as illustrated in Figure 7.3.

Examples of data rates and estimated bandwidth for robot sensors and control signals are reported in Table 7.1. Particular attention should be paid to the control loops,

Table 7.1 Requirements for robot sensors and signals [19].

Sensor name	Specs	Bandwidth
Cameras	2×, 640×480, 30 fps, 8/24 bit	147–441 Mb/s uncompressed
Microphones	2×, 44 kHz, 16 bit	1.4 Mb/s
F/T sensors	6×, 1k Hz, 8 bit	48 kb/s
Gyroscopes	12×, 100 Hz, 16 bit	19.2 kb/s
Tactile sensors	4000×, 50 Hz, 8 bit	1.6 Mb/s
Control commands	53DoF × 2–4 commands, 100 Hz/1 kHz, 16 bit	3.3 Mb/s (worst case), 170 kb/s (typical)

which cannot be executed locally. For instance, visual processing cannot be handled locally, because of the computational load/amount of data, and therefore is managed by a reasoning server remotely. This means a 5G wireless connectivity between peer points of Gb/s throughput, extremely low latency (< 5 ms) for force and motion control, and with failure rates below 10^{-7}, i.e. max 3.17s of outage per year. Also, for fully autonomous robots, to meet the required performance, it is very important to decide where and what to compute and how to transfer data and signals at a given transmission bandwidth [19].

7.3 5G Communication System

The 5G network architecture enabling the verticals was described in Chapter 1. This section presents the architecture, technology and security aspects of 5G that would cater for the Tele-healthcare use cases and applications presented in Section 7.2. The reader may find more information about 5G network domains, architectures and slicing in [22] and [23], respectively.

7.3.1 3GPP Technology Roadmap

The completion of the first phase (Phase 1 or Release 15, R15) of New Radio (NR) technology was in June 2018, in its Non-Standalone (NSA) Option 3 configuration. The Standalone (SA) Option 2 and 5 were finalized in September 2018. The remaining 3GPP NSA architecture options (4 and 7) will be specified in 2019. A high-level 3GPP content roadmap for 5G systems is depicted in Figure 7.4 [24].

The 3GPP R15 supports eMBB and some aspects of URLLC, such as: fast terminal processing time; front-loaded Demodulation Reference Signal (DMRS); mini-slot scheduling; uplink "grant-free" transmission; downlink pre-emption and flexible frame structure for low latency; downlink control channels with a high level of aggregation; data

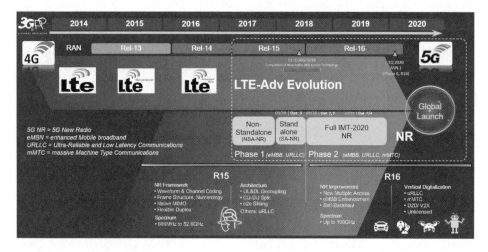

Figure 7.4 5G definition and 3GPP 5G content roadmap [24].

channel slot aggregation; and data (packet) duplication for high reliability. Some of those features are explained in Section 7.3.5.

The second phase (Phase 2 or Release 16, R16) of 5G will be frozen in 2020, and will support all usage scenarios: eMBB, URLLC and mMTC. Examples of features that will be standardized for further improving the reliability and latency performances are: Compact Downlink Control Information (DCI) and Physical Downlink Control Channel (PDCCH) repetition; Uplink Control Information (UCI) enhancements, such as Hybrid Automatic Repeat reQuest (HARQ) and Channel State Information (CSI) feedback improvements; Physical Uplink Shared Channel (PUSCH) enhancements, such as mini-slot level hopping, retransmission and repetition; scheduling, HARQ and CSI processing timeline enhancements; terminals multiplexing and prioritization for different latency and reliability requirements; enhanced UL configured grant transmissions (grant-free); multi-Tx/Rx Point (TRP) transmission; mobility and beam management improvements; uplink/downlink (UL/DL) intra-User Equipment (UE) prioritization and multiplexing. Some of the most important features to cater for Tele-healthcare use cases are described in Section 7.3.5.

7.3.2 5G Spectrum

The LTE operates always in the low band, i.e. below 6 GHz [24]. 5G NR is expected to support contiguous, non-contiguous and much broader channel bandwidths than available to current mobile systems. Also, 5G new radio will be the most flexible way to benefit from all available spectrum options from 400 MHz to 90 GHz, including licensed, shared access and licence-exempt bands, *Frequency Division Duplex* (FDD) and *Time Division Duplex* (TDD) modes, including *Supplementary Uplink* (SUL), Long-Term Evolution (LTE)/NR *uplink frequency sharing*, and narrowband and wide-band *Carrier Components* (CC). Operating band combinations for SUL are reported in [25]. More information on Dual Connectivity (DC) with and without SUL including UL sharing from terminal perspective – Output power dynamics for E-UTRA/NR DC (EN-DC) with UL sharing from UE perspective – may be found in [26].

A multi-layer spectrum approach is required to address such a wide range of usage scenarios and requirements [27]:

- The *Coverage and Capacity Layer* relies on spectrum in the 2 to 6 GHz range (e.g. C-band) to deliver the best compromise between capacity and coverage.
- The *Super Data Layer* relies on spectrum above 6 GHz (e.g. 24.25–29.5 and 37–43.5 GHz) to address specific use cases requiring extremely high data rates.
- The *Coverage Layer* exploits spectrum below 2 GHz (e.g. 700 MHz) providing wide area and deep indoor coverage.

5G networks will leverage the availability of spectrum from these three layers at the same time, and regulators are expected to make available contiguous spectrum in all layers in parallel, to the greatest extent possible.

7.3.3 5G Reference Architecture

The performance requirements of Section 7.2.1 and Section 7.2.2, especially in terms of reliability and end-to-end latency, cannot be entirely supported by 4G and/or Wi-Fi systems [1]. Superior wireless performance may be achieved with the deployment of a

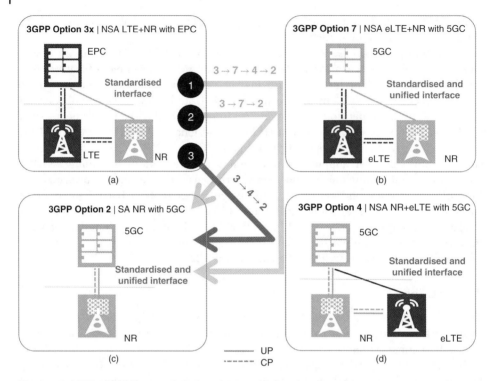

Figure 7.5 3GPP 5G architecture options and network migration strategies.

new generation of radio access networks with wireless access based on NR and (evolved) LTE (R15 LTE or eLTE of later releases). The most likely initial deployment options are illustrated in Figure 7.5 a) and c) (see e.g. [22] and [23]).

- *3GPP Option 3x* (NSA LTE plus NR with Evolved Packet Core, EPC) is the configuration that most carriers (network operators) will adopt, due to minor investments for their 5G launch. It supports eMBB and Fixed Wireless Access (FWA) usage scenarios and Voice over IP (VoIP) over LTE (VoLTE) or Circuit Switch Fall Back (CSFB) to earlier network releases (2G or 3G).
- *3GPP Option 2* (SA NR with 5G Core, 5GC) is initially adopted by only a few carriers globally. For taking full advantage from it, a wide coverage rollout is needed, as the interoperation with 4G/Evolved Packet System (EPS) is less efficient. Initial partial coverage rollouts may be more suitable for enterprise or overlay deployments. In the long run, it will support all usage scenarios (eMBB, URLLC, mMTC), plus other functionalities than Option 3x, such as Network Slicing and Voice over NR (VoNR).

The medium-/long-term migration path of 5G networks is also depicted in Figure 7.5, and, ultimately, all networks will converge to a 3GPP Option 2 architecture configuration (SA NR with 5GC). The medium-term migration strategies are basically two, depending on the carriers' spectrum availability for deploying the NR:

1. *From 3GPP Option 3x (NSA LTE + NR with EPC) to 3GPP Option 4 (NSA NR + eLTE with 5GC).* This choice is driven by the availability of low band NR (<3 GHz, <1 GHz for rural). The 5G services are launched with LTE+NR NSA on EPC, the NR and 5GC

rollout are driven by needs of 5G coverage; outside the NR coverage, 5G services may be provided by 3GPP LTE NSA Option 4 with 3GPP Option 5 (SA eLTE with 5GC). The interworking between eLTE and NR is also required.

2. *From 3GPP Option 3x (NSA LTE + NR with EPC) to 3GPP Option 7 (NSA eLTE + NR with 5GC).* The reasons to go for that are: Leverage 4G (LTE/EPC) installed base; NR rollout driven by better service (not coverage); and eLTE for all wide area coverage and all use cases. The drawbacks are: Full Dual Stack eNB/ng-eNB in LTE RAN to EPC/5GC; LTE RAN upgrades to eLTE; and required Interworking between LTE and NR. UE availability is also, currently, questionable.

As in previous mobile system generations, 3GPP defines a clear functional split between the NG-RAN and 5GC with the overall 5G System architecture defined in [28–30] and a more convenient overview of the Access Network (AN) and Core Network (CN) functions in [31]. The two network domains (AN–CN) are separated by a standardized interface (N2 and N3) defined in a set of specifications that enable multi-vendor RAN–CN deployments [29]. Also, this interface has been now unified, meaning that all next generation accesses (trusted/untrusted fixed/mobile 3GPP access points) must support this interface.

The 5GC is responsible for Non-Access Stratum (NAS) security and idle state mobility handling; UE internet protocol (IP) address allocation and protocol data unit (PDU) control; and mobility anchoring and PDU session management.

The NG-RAN is responsible for inter-cell Radio Resource Management (RRM), Radio Bearer (RB) control, connection mobility control, radio admission control, measurements configuration and provisioning, and dynamic resources allocation. As introduced in Chapter 1, and illustrated in Figure 7.6, the NG-RAN supports only two possible configurations:

- *Central Unit (CU)-Distributed Unit (DU) Split*: The RAN non-real-time protocol (NRT) stack is implemented in the CU and the functions more sensitive to delays in the DU close to the antennas.

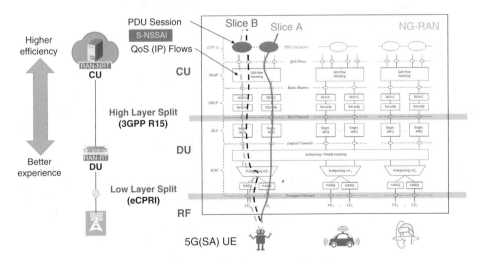

Figure 7.6 NG-RAN CU-DU and high layer and low layer protocols split.

- *CU-DU Co-Located at the Edge of the Network*: All RAN base band functionalities are running into one box placed closed to the antenna units.

The CU-DU split architecture can provide on-demand deployment [22, 23]: The CU may run on a cloud platform and can be deployed in different locations to meet the *efficiency-latency trade-off* for offering different type of services, e.g. the latency in regional, local and access data centre deployment scenarios is below 50 ms, 20 ms and 5 ms, respectively. Moreover, the cloud RAN architecture supports *multi-connectivity* and avoids data transmission detour, which reduces delays in user plane (UP) data transmission across the Multi-Radio Access Technologies (LTE-NR RATs). As further clarified in the following sections, multi-connectivity is critical for radio connection redundancy, and five uncorrelated links could already achieve an outage of less than 10^{-7}, which is the required hard-bound reliability for tactile applications [14].

In other words, in 5G, the radio interface protocol stack has been optimized for cloud and distributed computing with a flexible fronthaul split, between high layer protocols, i.e. Packet Data Convergence Protocol (PDCP) and Radio Link Control protocol (RLC), and low layer (L1) peer entities, using the enhanced Common Public Radio Interface (eCPRI). This will also allow access providers to relax the transmission capacity and the utilization of Ethernet. For example – assuming 100 MHz band, a three-sector site with 64Tx/Rx mMIMO and 16 layers – the interface between the radio unit (RU) and edge cloud (radio access unit, RAU) would require: 1 Tb/s with no split, using a CPRI interface; 150 Gb/s with low layer split and enhanced CPRI (eCPRI) interface; and 1–10 Gb/s with high layer split. The latency with low layer split is expected to be below 0.1 ms, as with CPRI; and 5 ms with high layer split, which is still satisfactory for supporting the WTS and WSR use cases described in Section 7.2 [23].

Additional benefits will come with the introduction of the 5GC, which supports many new enabling network technologies. Among other fundamental technology components, as depicted in Figure 7.7, the 5GC is characterized by a layered and

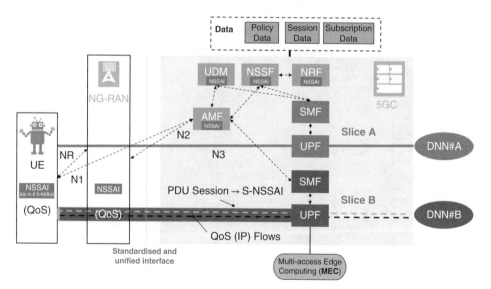

Figure 7.7 5GC and end-to-end network slicing [23].

service oriented architecture, with Control Plane (CP) and UP split and *Shared Data Layer (SDL)*, for subscription, state and policy data. It also supports: user plane session continuity when the terminal moves across different access points; interworking with untrusted non-3GPP access systems; a comprehensive policy framework for access traffic steering, switching and splitting; and wireless-wireline convergence [23].

The separation of control and user planes provides deployment flexibility and independence. The distribution of core functionality, especially user plane, closer to the radio, i.e. to the edge cloud, enables the placement of applications closer to the end user with improved throughput, reducing the transport network load, and latency, decreasing the physical distance and number of transport hops. The service-based architecture – including the related Network Repository Function (NRF) for 5GC control plane functions – allows flexible addition and extension of functions including the proprietary ones. The SDL, between all network functions (and micro-services within the functions), enables common session resiliency and geo-redundancy models, and a shared interface for accessing all session data (e.g. for analytics). The 5GC is also agile to deploy new or updated functions in a "DevOps" style. The Network Exposure Function (NEF) enables the integration of Telco functions with service providers, enterprises and operator's own IT systems to create new use cases and opportunities to monetize. The core slicing and related Network Slice Selection Function (NSSF) enable a flexible assignment of users to different network slice instances that are tailored to the different use cases. The 5GC also supports a unified subscriber management, authorization and authentication solution [23].

Other fundamental 5G enabling technologies, end to end, are: *Flow-based QoS*, with a much higher level of granularity than LTE, which is currently limited to the bearer service concept; multi-connectivity, where the robot could be connected simultaneously to 5G, LTE and Wi-Fi, offering a higher user data rate and a more reliable connection; terminal assisted Network Slicing, and E2E network management and orchestration, with in-built support for cloud implementation and edge computing [23].

The 5G flow-based QoS concept is illustrated in Figure 7.8. The NAS packet filters associate IP Flows with QoS Flows in the UE (robot) and User Plane Function (UPF). The Service Data Adaptation Protocol (SDAP) provides mapping between QoS flows and Data Radio Bearers (DRBs) and marking QoS Flow Identifier (QFI) in both DL and UL packets. There is a single SDAP entity for each PDU session (GTP Tunnel). The QFI uniquely identifies the QoS flow and consists of the following parameters [29]:

- 5G QoS Identifier (5QI): Resource Type (Guaranteed Bit Rare (GBR) and Non- GBR), Priority Level, Packet Delay Budget (PDB) UE ↔ UPF, and Packet Error Rate (PER);
- Allocation Retention Priority (ARP);
- Guaranteed (GFBR) and Max Flow Bitrate (MFBR) for GBR flows (UL and DL).

In the DL, each UPF uses policies from Policy Control Function (PCF)/Session Management Function (SMF) to identify flows and adds QFI tags, enforces Session-AMBR (Aggregated) and counts packets for charging. Each UPF sets Reflective QoS (RQI) to activate Reflective QoS in the UE (robot). The NG-RAN uses QFI tags and policies to map flows to one or more DRBs, and enforces the Max Bit Rate (UE-AMBR) limit per UE (robot) for non-GBR QoS flows.

In the UL, the UE (robot) uses either signalling or "reflective" learning approach to derive the QFI policies to map QoS flows onto DRBs, and performs UL rate limitation on

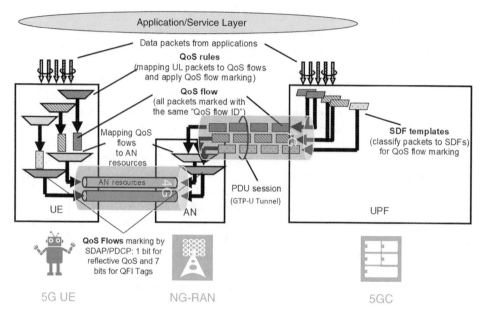

Figure 7.8 5G QoS Concept [29].

a PDU Session basis for non-GBR traffic, based on Session-AMBR value. The NG-RAN and UPF police QFI usage and enforce Max Bit Rate (UE-AMBR) and Session-AMBR, respectively. The UPF counts the packets for charging.

The transport network layer QoS treats packets based on 5QI tags (e.g. using DiffServ). The mapping is based on Software Defined Networking (SDN) policies.

IP Flows are mapped onto QoS Flows, which are mapped onto one or more DRBs. Data radio bearers are associated to one PDU Session, which is mapped onto one S-NSSAI. The S-NSSAI is mapped onto one Network Slice Instance (NSI), i.e. one Network Slice; and the NSI is mapped onto a single Data Network Name (DNN). However, the vice versa is not valid, as described in the following paragraphs. This is how 5G handles the flow-based QoS within a given NSI [23].

3GPP for terminal assisted network slicing defines a new parameter denoted as *Single-Network Slice Selection Assistance Information* (S-NSSAI). Each S-NSSAI assists the network in selecting a network slice instance. The S-NSSAI is composed by the following attributes [29, 30]:

- *Slice/Service Type (SST)*: Where 1 (eMBB), 2 (URLLC), 3 (mMTC) are the standardized values for roaming; operator-specific settings are also possible;
- A *Slice Differentiator (SD)*: Tenant ID, optional, for further differentiation during the NSI selection.

The Network Slice Selection Assistance Information (NSSAI) consists of a collection of S-NSSAIs. In 3GPP R15, a maximum of eight S-NSSAIs may be sent in signalling messages between the UE (robot) and the Network.

The NSSAI is configured (Configured NSSAI) in the UE (robot) per Public Land Mobile Network (PLMN) by the Home PLMN (HPLMN). The robot uses the Requested NSSAI during the Registration Procedure and the Allowed NSSAI, received from the

Access and Mobility Function (AMF), within its Registration Area (RA). The RA allocated by AMF to the robot has homogeneous support for network slices [29].

The 5GC supports AMF-level slicing per terminal type, and SMF- and UPF-level slicing per service or per tenant based on S-NSSAI and DNN. An example of two network slices for one terminal is illustrated in Figure 7.7.

The NG-RAN is aware of the slice at PDU Session level, because the S-NSSAI is included in any signalling message containing PDU Session info. Pre-configured slice enabling in terms of NG-RAN functions is implementation-dependent. An example of NG-RAN slicing is depicted in Figure 7.6. The Medium Access Control (MAC) scheduling – based on RRM policies related to the SLA in place for the supported slice and QoS differentiation within the slice – is vendor-dependent [29].

Details on how the NSSAI is exchanged between the terminal and network entities during the Registration and PDU Session establishment procedure, as well as how the NSI is instantiated end-to-end, may be found in [23].

7.3.4 5G Security Aspects

5G System security is based on the well-established and proven 4G/EPS security, which has been further enhanced [32, 33]. The 5GS keying hierarchy is comparable to 4G for the functionality towards the RAN, i.e. all keys for the *Access Stratum* (AS) are derived from the NAS security parameters inside the Core Network and signalled the RAN. The main new model of the 5GS is on how the security functionality is decomposed and distributed inside the core network. This also enables the globally unique 5G *Subscription Permanent Identifier* (SUPI, which is comparable to the IMSI of earlier system generations) to always be signalled encrypted via the RAN towards the CN. It is decrypted by the Home PLMN and delivered from there to the serving Core Network for any user service, management and regulatory purposes. In contrast to earlier system generations, where the International Mobile Subscriber Identity (IMSI) was used in the RAN for recovering from network failures and thereby enabled certain attacks, the 5G System never exposes the SUPI to the RAN nor does it transfer it in clear via the radio. Further, 3GPP 5G Release 15 adds an option to perform user plane integrity protection between UE (robot) and gNB. In 3GPP Release 16, security algorithms use up to 256-bit keys [32] (see Figure 7.9).

Also, the transport network layer within and between RAN and core network domains is protected using IPSec tunnels. Examples of security deployment scenarios for 3GPP NSA Option 3x and Option 2 configurations are illustrated in Figures 7.10 and 7.11, respectively. As shown in the figures, the 3GPP Option 2 has the same level of security as 3GPP Option 3x. Hence, if 4G is proved to be an isolated network, 5G results even more robust, and meets all security requirements to connect robotic platforms. More information on 5G security aspects and deployment scenarios may be found in [34].

7.3.5 5G Enabling Technologies

This section summarizes key contributions of researchers from industry and academia that address the technical challenges reported in Section 7.2 and sheds light on some fundamental 5G technologies for URLLC, and especially those suitable for Tele-healthcare applications. More information on the subject may be found in [35].

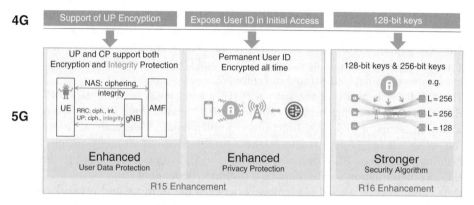

Figure 7.9 3GPP 5G security enhancements [32].

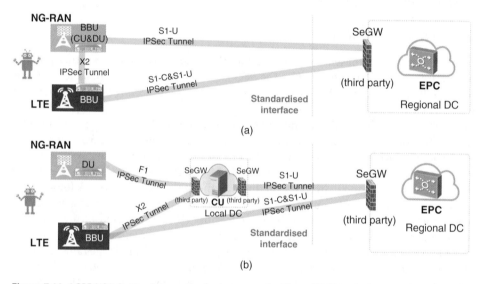

Figure 7.10 3GPP NSA Option 3x security deployment: a) without CU-DU split; b) with CU-DU split.

7.3.5.1 5G Design for Low-Latency Transmission

In 3GPP R15, and later releases, both NR (TDD and FDD) and LTE (TDD) are designed to fulfil the requirements on latency reported in Section 7.2.

LTE and NR use orthogonal frequency division multiplexing (OFDM) as the waveform. A Cyclic Prefix (CP) is used to cope with a time-dispersive propagation channel, i.e. only delay-spread of signals that exceeds the CP length introduces inter-symbol interference. In the UL, LTE uses discrete Fourier transform (DFT) precoding to reduce the peak-to-average-power ratio at the transmitter, and DFT precoding is also available for NR uplink [36].

LTE uses a fixed numerology of 15 kHz sub-carrier spacing and operates below 6 GHz. The NR has been designed for all spectrum options and supports a *flexible numerology*, which consists of different Sub-Carrier Spacing (SCS), nominal Cyclic Prefix (nCP) and Transmission Time Interval (TTI), or scheduling interval, depending on bandwidth and latency requirements. Sub-carrier spacing of 15 kHz to 120 kHz, and the corresponding

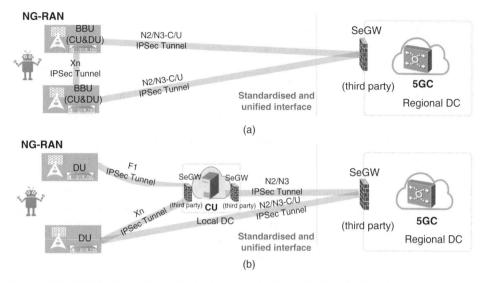

Figure 7.11 3GPP SA Option 2 security deployment: a) without CU-DU split; b) with CU-DU split.

cyclic prefix of 4.7 to 0.6 μs and scheduling interval of 1 ms to 0.125 ms, are defined for different Carrier Components (CC), which may vary from 5 MHz to 100MHz, below 6GHz, and from 50 to 400MHz, above 6GHz [23]. Examples of OFDM numerology options are illustrated in Figure 7.12.

For optimal radio performance, the higher the carrier frequency, the higher the allowed carrier component and sub-carrier spacing, the lower the corresponding cyclic

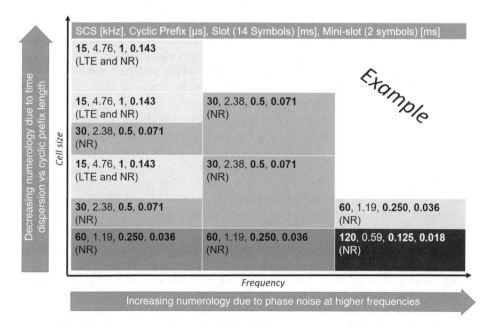

Figure 7.12 Examples of feasible OFDM numerology options for different spectrum ranges and cell deployments [36].

prefix and scheduling interval. At higher frequencies (e.g., millimetre-wave, or spectrum above 20 GHz), the phase noise increases and numerologies with larger sub-carrier spacing provide better robustness. This is because at higher SCS the symbol duration decreases, and hence also the length of a *slot*, which is the basic frame structure at which most physical channels and signals repeat. In LTE and NR (R15) a slot comprises 14 OFDM symbols (OFDMS), which leads to a slot length of 1 ms at 15 kHz SCS. In NR, by using higher numerologies, the slot duration decreases from 1 ms to 0.125 ms, which is beneficial to achieving lower latencies [36].

In NR, beyond numerology scaling, the concept of *non-slot-based transmission* has been introduced, which is also referred to as *mini-slot*. A mini-slot in NR can start at any OFDM symbol and can have a variable length of 2, 4, or 7 symbols [37]. This provides fast transmission opportunities for URLLC traffic, because it is not restricted to any particular slot boundary. Thus, the mini-slot provides a viable solution to low-latency transmissions irrespective of sub-carrier spacing, while the usage of wider sub-carrier spacing for low latency is mostly limited to small cells deployment [36].

Similarly for LTE FDD spectrum allocations, the *short TTI* (sTTI) enables fast transmission opportunities of 2, 3 or 7 OFDMS duration. The sTTIs can be embedded into the existing LTE frame structure, around existing control channels and reference symbols, and are independently scheduled with new in-band control elements, thereby allowing lower scheduling delay at the expense of increased overhead [36].

The NR supports both *static and dynamic UL/DL TDD configuration options*. The TDD latency is largely determined by the worst case timing; for example, a DL packet arrives exactly at the beginning of an UL allocation or vice versa. A suitable TDD configuration for URLLC is when UL (mini-)slots alternate with DL (mini-)slots. However, mini-slots shorter than 7 OFDM symbols are not considered for TDD, as the UL-DL switching overhead would become dominant in too-short time slots [36].

Furthermore, for the UL, a significant part of delay is due to the lengthy time it takes while exchanging scheduling requests and UL grant transmissions between the BS and the UE (robot). Also, for a reliable communication this signalling bears the risk that both messages need to be correctly decoded in order to start an UL transmission. Many "grant-free" multiple access techniques that eliminate the dynamic request and grant signalling overhead have been investigated. For instance, as a remedy for both the delay and robustness issues, a *periodic grant* can be configured in the UE (robot). Both LTE and NR support a Semi-Persistent Scheduling (SPS) framework, in which the terminal is given a periodic grant that it uses only when it has UL data to transmit. Necessarily, these URLLC resources are tied up for the UE (robot), but by assigning overlapping grants to multiple UEs the resource waste is reduced, and at lower rates the impact on reliability due to collisions can be managed. This method of periodic grants reduces the latency by one feedback round-trip time and thereby enables low-latency UL data transmission [36].

A significant part of the delay budget relates to the *processing time for decoding in the DL*. The usage of Low-Density Parity Check (LDPC) codes for 5G NR eMBB services has the promise of reducing it. However, the effective benefits of such codes for URLLC services are still disputed in 3GPP. To further reduce the processing time for HARQ retransmissions, it is proposed to *predict whether the decoding will be successful or not*; this enables the UE (robot) to anticipate its feedback transmission while running the decoding in its pipeline. False positives, which occur when an early ACK is generated for a transport block that is not going to be correctly decoded, are considered more critical than false negative, because they may trigger a higher layer retransmission. Prediction techniques with improved reliability are currently for further study. Improved prediction techniques with sensitive tolerance to false negatives will most probably cope with the Tele-healthcare targets [38].

A resource- and latency-efficient scheduling solution is to multiplex data with different TTI lengths (i.e., mini-slots and slots) and, if needed, let the high-priority data use resources of lower-priority data. This type of multiplexing is also referred to as *pre-emption*. For example, in NR DL, a mini-slot carrying high-priority or delay-sensitive data can pre-empt an already ongoing slot-based transmission on the first available OFDM symbols, without waiting until the next free transmission resource. A similar concept is also considered for the UL, and in general for LTE. At the cost of degrading the longer transmission, no additional resources would need to be reserved in advance for the low-latency communication. The damaged longer transmission is then swiftly repaired with a transmission containing a subset of the Code Block Groups (CBGs) in a later TTI, after providing the essential information to clean the contaminated soft values in the receive buffer from the pre-empted data [36].

Massive Multiple-Input Multiple-Output (mMIMO) is an integral part of the 5G NR features from day one. With mMIMO the number of transmitting antenna elements is much higher than the number of MIMO streams (layers). In practice, mMIMO means that the number of controllable antenna elements is more than eight. 5G NR supports 8 Layer Single User (SU)-MIMO or 16 Layer Multi-User (MU)-MIMO in DL, and 4 Layer SU-MIMO in UL, with the possibility of dynamic switching in both directions. MU-MIMO means that parallel MIMO data streams (or layers) are transmitted to different robots at the same time-frequency resources.

Beamforming (BF) offers the advantages that the same resources can be reused for multiple robots in a cell. It allows *Space Division Multiple Access* (SDMA), maximizing the number of supported robotic platforms within that sector. Also, it minimizes interference, increases the cell capacity, enhances link performance and increases the coverage area. 5G radio design is fully optimized for mMIMO using three basic techniques for forming and steering beams [23]:

- *Digital Beamforming*, where each antenna element has a transceiver unit with the adaptive Tx/Rx weights in the baseband, enabling frequency-selective beamforming. Digital beamforming boosts capacity and flexibility and is mostly suited to bands below 6 GHz.
- *Analogue Beamforming* implements only one transceiver unit and one RF beam per polarization. Adaptive Tx/Rx weighting on the RF is used to form a beam. This is best suited for coverage at higher mmWave bands and offers low cost and complexity.
- *Hybrid Beamforming* is a combination of analogue and digital beamforming. When some beamforming is in the analogue domain, the number of transceivers is typically much lower than the number of physical antennas, which can simplify implementation, particularly at high frequency bands. This technique is suited to bands above 6 GHz.

In NR, all physical control and data channels support beamforming. The many radiating dipoles allow narrower beams in the zenith (vertical plane) and azimuth (horizontal

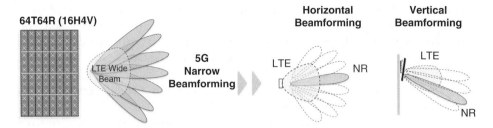

Figure 7.13 5G Massive MIMO narrow beam design for 3D shaping.

plane) directions, as depicted in Figure 7.13. The coverage for common and shared channels may be thus improved, e.g. the coverage gain of 64T64R vs. 2T2R may be up to 10 dB, and the interference can be also mitigated, yielding a higher Signal to Interference plus Noise Ratio (SINR), meaning better throughput and more cell capacity.

Multi-Access Edge Computing provides efficient transport and low latency to hosted applications in the operator trusted domain, close to the robot's access point (edge). As shown in Figures 7.3 and 7.7, the MEC enables access to applications deployed near the robotic platform [21]. In 5G, the local break-out (UPF of 5GC) could be flexibly deployed in other locations too, leveraging Network Function Virtualisation (NFV) and End-to-End Service Orchestration (E2E SO) capabilities, i.e. algorithms within the different domains (RAN, Core, Transport etc.) that enable the right flexibility and trade-offs for the network administrator to efficiently place VNFs and exploit slicing [23].

7.3.5.2 5G Design for Higher-Reliability Transmission

The definition of reliability and related performance targets for the WTS and WSR use cases were presented in Section 7.2. In the following, we review the key disruptive innovations that enable a reliable connection for robotics control and beyond.

The first disruptive innovation is about a more flexible design of *control channels*, specified in earlier 3GPP releases to support all service applications and now tailored to an individual service. Examples of specific control channel enhancements are [38]:

- *Link Adaptation for Control Channels*: In LTE, the Physical Downlink Control Channel (PDCCH) already supports link adaptation by adjusting the *aggregation level* (i.e., repetition coding rate) in accordance with radio conditions. The maximum supported aggregation level is 8, which provides a PDCCH decoding error rate of 1% for an SINR of −5 dB. For 5G, with larger aggregation levels (e.g., 16), more robust coding schemes would be required to achieve much lower error rates. A similar approach should be applied to other critical control channels, such as the Physical Uplink Control Channel (PUCCH), carrying the channel quality indicator (CQI) and HARQ feedback. Besides varying the coding rate, *power control techniques* could be also applied to boost or reduce the transmit power on a per-terminal basis and achieve the required error performance.
- *Asymmetric Signal Detection for ACK/NACK*: For URLLC, the reliability of detecting NACK signals is more important than the trustworthiness of detecting ACK signals. Therefore, the asymmetric signal detection can be utilized to reduce the probability of

missing a NACK signal at the cost of increasing a wrong ACK detection (i.e., decoding an ACK as a NACK).

- *Compact Downlink Control Information (DCI)*: URLLC services are typically associated with smaller packet sizes compared to eMBB. Hence, the number of bits that are used to indicate the number of Physical Resource Blocks (PRBs) can be reduced. For 5G NR, the DCI content is still under discussion, and two DCI sizes have been agreed for the purpose of performance evaluation: 20 bits and 60 bits. The compact DCI formats allow more robust transmission of the control channel information without increasing the control overhead.
- *CQI Enhancements*: The CQI report is utilized for performing the downlink MCS adaptation by indicating to the Base Station (BS) the highest MCS that the terminal (robot) estimates it can decode at a certain BLER level. The following three enhancements for the CQI measuring and reporting procedure are proposed. First, the CQI report must be estimated at the UE (robot) targeting different Block Error Rate (BLER) targets compared to the fixed 10% target in LTE. Second, the BLER target for DL data transmission should be configurable according to the TTI duration, HARQ round-trip time (RTT) and control channel reliability. Third, the CQI report should effectively reduce the link adaptation mismatches due to the rapid and unpredictable SINR variations.

In *multi-connectivity*, a robot is connected to the network via multiple radio links. Several flavours of multi-connectivity have been defined in 3GPP for LTE and NR to improve user throughput and transmission reliability by aggregating signals of different access points. For instance:

- In R10, 3GPP introduced *Carrier Aggregation* (CA) as a method for the device to connect via multiple carriers to a single base station. In CA, the aggregation point is the MAC entity, allowing a centralized scheduler to distribute packets and allocate resources among all carriers, e.g. based on propagation channels knowledge, with a tight integration of the radio interface protocols involved.
- In R12, *Dual Connectivity* (DC) was specified for LTE, where the aggregation point was moved up to the PDCP layer. This way, two MAC protocols with their separate scheduling entities can be executed in two distinct nodes, without strict requirements on their interconnection.
- In R15, both architecture concepts of CA and DC are reused in both LTE and NR to improve the transmission reliability on upper protocol layers, beyond reliability improvements intended on the physical layer (PHY). This is achieved by *Packet Duplication* (PD) at PDCP layer. An incoming packet, for example, of a URLLC service, is thereby duplicated on the PDCP level, and each duplicate undergoes procedures on the lower layer protocols, RLC and MAC, and hence individually benefits from retransmission reliability schemes on RLC, MAC and PHY. Eventually, the data packet is transmitted via different frequency carriers to the robot, which ensures uncorrelated transmissions due to frequency diversity and even macro diversity, in DC from different sites. The method is illustrated in Figure 7.14 for both CA and DC [36, 38].

Furthermore, multi-connectivity based on dual connectivity has the potential to achieve zero *Mobility Interruption Time* (MIT) and zero *Handover Failure Rate* (HFR), i.e. *Mobility 0-0*, for user plane data and signalling messages. The handover is made

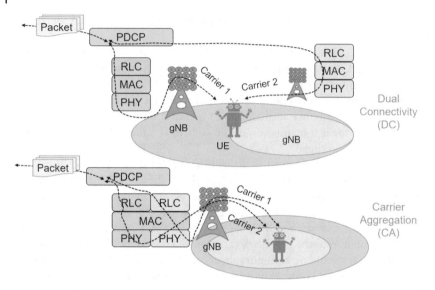

Figure 7.14 Packet duplication in DC and CA; conceptually applicable to LTE and NR [36, 39].

in two steps and during the procedure packet duplication may be also employed [40, 41]:

- *Make before break and RACH-less handover:* In R14, with a single transceiver (TRX), simultaneous transmission and reception is not possible, hence the device disconnects from the Serving Cell (SC) after receiving a Handover Command (HO CMD) and resumes only when it receives the first packet from the Target Cell (TC). The Random Access (RA) procedure can be avoided if the Time Advance (TA) is null or the same for both source and target cells, and an UL grant is sent to the device in an HO CMD, or the terminal receives it on the PDCCH from the target node (T-gNB). Using this technique, the Radio Link Failure (RLF) occurs frequently, due to interference, and the device stops transmitting packets. The improvements range from 40 ms to 5 ms, but only under certain circumstances.
- *DC-based handover:* In R15, and later releases, the device with multiple (two) TRXs can be simultaneously connected to the source and target cell. The source is removed only if the signal falls more than an additional threshold (offset) below the currently best configured gNB; the master role is swapped before the source gNB is removed from the active set; and the signalling radio bearer (SRB), i.e. control plane messages carrying uplink measurement reports and downlink reconfigurations, is established and duplicated via both master node (MN) and secondary node (SN). The NR RLF detection reuses the same mechanism as with LTE; it could be extended with a feature for "SN survival", reporting the failure of the MN via the SN and swapping the MN. If this solution was in place, the MIT and HFR would reduce to 0-0.

7.3.6 5G Deployment Scenarios

The 5G deployment scenarios using a 3GPP NSA/SA architecture configuration is depicted in Figure 7.15. All network domains, including some functionality of the RAN, may run on cloud infrastructures.

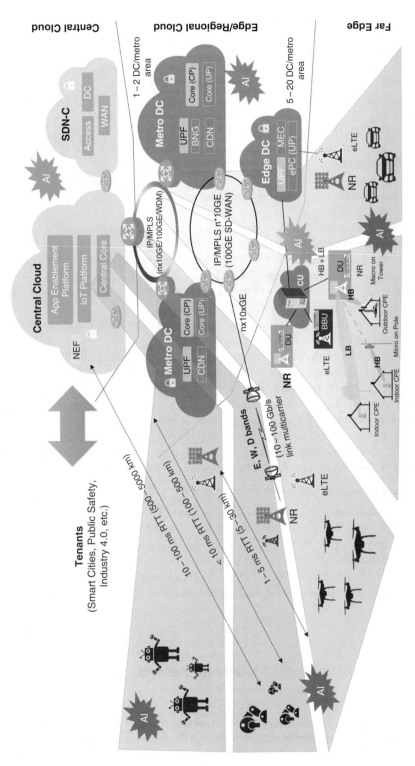

Figure 7.15 Example of 5G deployment scenario on EPC/5GC with end-to-end network slicing.

The *far edge* data centre, or point of presence (PoP), may host CU functions, as illustrated in Figure 7.6. This is the area where radio equipment (antennas, radio and base band units) are deployed.

The *edge/regional* cloud is separated from the far edge zone by a standardized (NSA Option 3x) and unified (SA Option 2/5, NSA Option 4/7) interface, see Figures 7.5 and 7.7, leaving a clear division between radio and core domains. The local break-out to a MEC server (see Figure 7.7) may be located in an edge/regional PoP.

The core network functionalities may be placed in the edge/regional and *central* clouds. The IoT and application enablement platforms reside in the central cloud, where the SDN controller of the transport network layer are also located.

The introduction of the 5G core will be based on software upgrades of the core functions instantiated in the edge/regional cloud, namely in the metro and edge data centres, as shown in Figure 7.15.

The utilization of network slices – five in Figure 7.15, with different service level agreements between tenants and Communication Service Providers (CSPs), e.g. in terms of *throughput, reliability, latency* and *mobility* – will allow the infrastructure to support a number of use cases – five in Figure 7.15, i.e. 5G for WSR; 5G for WTS; 5G for Augmented Reality (AR), Virtual Reality (VR) and Mixed Reality (MR); 5G FWA; and 5G for connected cars – over a single physical infrastructure.

In the case of Tele-healthcare, this partitioning will make the 5G business case positive for the CSP and tenants, as discussed in Section 7.4.

The expected delays between robots and far/edge (5–30 km), edge/regional (100–500 km) and central cloud (500–5000 km) are 1–5 ms, below 10 ms and 10–100 ms, respectively, as shown in Figure 7.15. Based on this, the CSP may decide the optimal location of network functions and MEC servers to meet the requirements summarized in Section 7.2 for WTS (see Figure 7.2) and WSR (Figure 7.3).

Artificial intelligence, e.g. algorithms for descriptive, predictive, prescriptive and cognitive analytics; deep learning; reasoning etc. will find application mostly in the following equipment, domain and functions, but not limited to those [23]:

- Self-Organizing Network (SON): Key capabilities, algorithms and architecture attributes within the different domains (RAN, Core, Transport etc.);
- data and application enablement platforms, i.e. big data analytics (structured data analytics, text analytics, web analytics, multimedia analytics, network analytics, mobile analytics, etc.);
- platforms for IoT and Customer Experience Management (CEM); and
- robotic platforms, vehicles, Customer Premise Equipment (CPE) and next generation devices.

7.4 Value Chain, Business Model and Business Case Calculation

In this section, we qualify and quantify the impact of robotics globally, and present an example of business model for Tele-healthcare and the corresponding business case calculation for the communication service provider and care service organization. More information on the topic may be found in [4].

7.4.1 Market Uptake for Robotic Platforms

In 2014, more than 1000 robotic platforms were sold for medical applications, primarily (about 80%) for remote-assisted surgery. Ten times more medical units have been forecast to be sold by the end of 2019 [42]. In the future, safety, better clinical outcomes and especially reduced labour costs will lead to exponential growth in demand not only for robot-assisted surgery, but in other segments of healthcare as well, such as sanitation, sterilization, lab processing and materials handling.

The presence of service robots is currently insignificant at the global level, but sales of robotic platforms to assist elderly and disabled persons will exceed 12,000 units by end of 2018 [42]. This market is expected to increase substantially over the next 20 years. In fact, looking into the future, we are likely to see hospitals, nursing homes and hospices, as well as private houses for the elderly, in developed and emerging economies, as key drivers for the widespread use and demand for wireless service robots.

By 2030, 16% of the world population will be over 60 years of age [17]. The speed of growth of ageing population is much more dramatic in developed countries. For example, in Japan, it is anticipated that by 2050, 39% of all Japanese will be above 65 years old. Under such conditions, the need of service robots for elderly care, such as physical and preventive care, as well as companionship will rapidly surge. In Australia, health also dominates public expenditure and employment. Australia spends more than 10% of GDP ($AU170.4 billion) on health, making it a prime candidate for innovation to both reduce costs of healthcare and improve outcomes [43].

As a result, the global market of personal robots, including "care-bots", driven by rapidly ageing populations, could reach up to €14.4 billion by 2022 [42]. The global total sales revenue of household and personal service robots is forecast to grow by 23.5% per annum in the 2015–2020 period. Personal care robots, priced US$1500–US$4700, could enter our personal lives, commercializing the market by 2020.

7.4.2 Business Model and Value Chain

Traditional operators will not likely offer WTS or WSR services as standalone providers. Given the required level of specialization, understanding of professionals' specific needs, an operator will need to partner with specialized sectors in healthcare, either with clinic/medical centres or suppliers of robotic platforms. Suppliers of medical robots include accessories and services, and provide leasing contracts for their robots. The operator will essentially guarantee connectivity in strict SLA with WTS/WSR providers, where the latter may be a medical centre (custom application) or a separate third party, who can provide services to multiple medical centres/clinics. The connectivity provider could adopt new tariff models by selling guaranteed URLLC services to enable new revenue models on top of only bandwidth being sold.

In our model, the value chain consists of: *third-party robot provider/operator ↔ medical clinic centre ↔ specific department ↔ patient*. The contribution of an insurance company for monetization and flow can be either through patients or directly to department in the medical centre or the medical centre itself.

7.4.3 Business Case for Service Providers

The proposed business model reflects the financial impact on a) a Mobile Network Operator to build-up a 5G Indoor Infrastructure (ID); and b) a Care Provider to deliver "robot care" services.

7.4.3.1 Assumptions

For the operator business case calculation, we assume the 5G indoor infrastructure (small cells) is deployed in hospices, residential areas and private buildings, under an existing macro coverage, e.g. LTE services. The operator's costs consist of CAPEX and related OPEX for ID deployment and backend servers. The revenue is estimated based on ARPU per robot connectivity (€50/month).

In the operator case, the Present Value (PV) of the cash flow is derived over 10 years – from 2020, when the 5G mobile system 3GPP R15 is launched to 2029 – assuming a Discount Rate (PV) of 8%, an Earnings Before Interest, Taxes, Depreciation and Amortization (EBIDTA) margin of 33%, an ARPU decline multiplier of 2%, an equipment price erosion multiplier of 8%, and an OPEX erosion multiplier of 1%.

The number of service robots and deployed indoor cells for connecting them to the backend server are reported in Table 7.2. From 2020 to 2029, the number of service robots in hospices (H), residential floors (R) and private homes (P) is expected to grow year on year by 12%, 10% and 33%, respectively.

In the case of a Care Provider, the corresponding cash flow (PV) in 2020–29 is derived assuming a Discount Rate (PV) of 8% and an EBIDTA Margin of 40%.

Other relevant parameters for the Care Provider business case calculation are reported in Table 7.3.

7.4.3.2 Business Cases Calculation

In the period 2020–29, the cash flow (PV) for the Connectivity Service Provider (Operator) and for the Care Provider, under the assumptions summarized in Section 7.4.3.1, are depicted in Figures 7.16 and 7.17, respectively.

Figure 7.16 shows an immediate Return on Investment (RoI) for the Operator: The cumulative value of the Net Present Value (NPV) of cash flow remains flat during the first year; it then soars proportional to the increase of the number of deployed service robots year on year till 2029.

The business case for the Care Provider is different, and the breakeven point proves to be in 2024, i.e. four years after the "Robot Care" launch in 2020. Then, the cumulative value of the cash flow (NPV) becomes positive and keeps rising steadily till 2029, where it reaches €2.636 million.

Although not shown in Figure 7.17, it is worth noting that the cumulative value of the cash flow (NPV) for the Care Service Provider is truly sensitive to the CAPEX per robot; e.g. if the price of each robot was set to €2500, i.e. 25% more expensive than the value reported in Table 7.3, and used for the business case calculation, the breakeven point would be in 2026, i.e. two years later!

Table 7.2 Number of service robots and indoor (ID) cells per location in 2020 [4].

Parameter	Hospices (H)	Residential floors (R)	Private homes (P)
Robots	500 (H)	200 (R)	100 (P)
ID Cells	H/20	R/10	P

Table 7.3 Assumptions for the care provider business case calculation [4].

Parameters	Value
Useful time of deployment per day [h]	12
Number of visitations, hospice, per day [#]	12
Number of visitations, residential type, per day [#]	6
Robot replacement every Nth year [#]	5
Backend servers per robot [#]	0.05
Costs	
CAPEX per robot [€]	**2,000**
OPEX per robot [€/y]	2,443
CAPEX backend server [€]	5,000
OPEX backend server [€/y]	2,500
Connectivity [€/month]	**50**
Maintenance [€/month]	75
Insurance [€/month]	75
Electricity consumption [kW/h]	65
Electricity cost [€/kWh]	0.15
Revenues	
Hospice robot, shared [€/month]	90
Residential robot, shared [€/month]	150
Private robot [€/month]	900

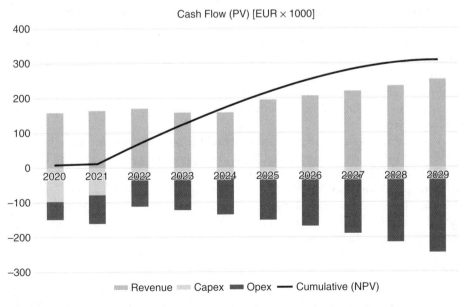

Figure 7.16 Cumulative cash flow (Net Present Value) for Connectivity Service Provider.

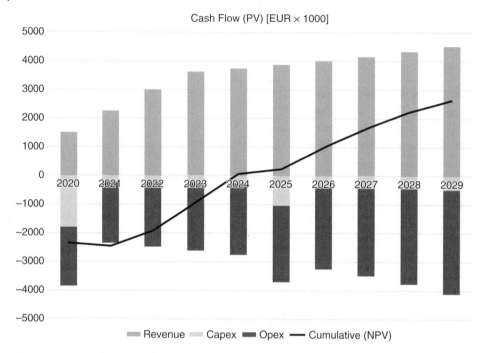

Figure 7.17 Cumulative cash flow (Net Present Value) for Robot Care Service Provider.

7.5 Conclusions

In this chapter, we have presented two examples of use cases for healthcare using robotic platforms enabled by ultra-reliable low-latency communications, which require the deployment of 5G wireless.

It should be noted that the proposed 5G wireless solutions to Wireless Tele Surgery and Wireless Service Robots, for Tele-health and Tele-care services, are applicable to many more use cases and scenarios, because the *haptic control* is inherent to a number of tactile applications, and *control loops* between reasoning servers and machines are needed for handling traffic of connected vehicles, robots for industrial processes automation, etc. [43].

As a part of this framework, we have analysed the business case for a Connectivity Provider and for a Care Provider using WSR in hospitals, hospices and related campus areas and, especially, at home (private buildings).

A more attractive business case for operators would require to support other use cases using the same physical infrastructure, in order to cover costs (especially OPEX) of a dedicated 5G system to Tele-healthcare. Network slicing could be a fundamental enabling technology to this end.

For Care Providers, the business case is quite sensitive to rising OPEX and CAPEX per robot, including its replacement. The business case will unquestionably fly when more powerful robotic platforms will help save more costs of care, and their price will be much more accessible to consumers. Business to Business (B2B) is where a RoI business would be easily made, as long as the price per month to lease the robot was much cheaper than hiring people, as confirmed in [44].

Ultimately, we want to state clearly that the assumptions and views reported herein are solely those of the authors, and do not necessarily represent those of their affiliates.

References

1 NGMN Alliance. NGMN 5G white paper. February, 2015.
2 ITU-T Rec. Y.3101. Requirements of the IMT-2020 network. January, 2018.
3 3rd Generation Partnership Project. Study on scenarios and requirements for next generation access technologies. 3GPP TR 38.913, v.14.3.0.
4 D. Soldani, et al. 5G mobile systems for healthcare. *Vehicular Technology Conference (VTC Spring)*, Sydney, June 4– 7, 2017.
5 J. Marescaux, J. Leroy, F. Rubino, M. Smith, M. Vix, M. Simone, D. Mutter. Transcontinental robot-assisted remote telesurgery: Feasibility and potential applications. *Advances in Surgical Technique*, 235(4):487–492, 2002.
6 S. E. Butner, M. Ghodoussi. Transforming a surgical robot for human telesurgery. *IEEE Transactions on Robotics and Automation*, 19(5):818–824, 2003.
7 S. Avgousti, et al. Medical telerobotic systems: current status and future trends. *BioMedical Engineering OnLine*, 15(1):96, 2016.
8 Rysavy Research. Mobile Broadband Explosion: the 3GPP wireless evolution. Aug. 2012. http://www.4gamericas.com/en/.
9 G.H. Ballantyne. Robotic surgery, telerobotic surgery, telepresence, and telementoring – Review of early clinical results. *Surgical Endoscopy*, 16(10):1389–402, 2002.
10 S. J. Baek, S.H. Kim. Robotics in general surgery: an evidence-based review. *Asian J Endoscopy Surgery*, 7(2):117–23, 2014.
11 D. Soldani, R. Tafazolli. 5G: The nervous system of the Silver Economy. *IEEE MMTC E-Letter*, 10(4):23–26, 2015.
12 C. Jay, M. Glencross, and R. Hubbold. Modelling the effects of delayed haptic and visual feedback in a collaborative virtual environment. *ACM TOCHI*, 14(2): article no 8, 2007.
13 A. B. Weddle, H. Yu, M. Moreno. UX impacts of haptic latency in automotive interfaces. Review and User Study, *Imm. Corp.*, 2013.
14 M. Simsek, A. Aijaz, M. Dohler, J. Sachs, and G. Fettweis. 5G-enabled tactile Internet. *IEEE Journal on Selected Areas in Communications*, 34(3): 460–473, 2016.
15 Y. Makino, Y. Furuyama, S. Inoue, H. Shinoda. HaptoClone for mutual tele-environment by real-time 3D image transfer with midair force feedback. *Proceedings of CHI Conference on Human Factors in Computing Systems*, pages 1980–1990, USA, May 7–12, 2016.
16 P. Dario, et al. Ambient Assisted Living Roadmap. September, 2014.
17 Merrill Lynch, Bank of America. The Silver Dollar – Longevity Revolution Primer. Report, Jun. 2014.
18 European Commission. *The Silver Economy*. Final Report, ISBN 978-92-79-76911-5, 2018.
19 G. Metta. Robotics-derived requirements for the Internet of Things in the 5G context. *IEEE MMTC E-Letter*, 9(5):13–16, 2014.

20 F. Cavallo. Toward cloud service robotics in active and healthy ageing applications. *IEEE MMTC E-Letter*, 10(4):31–34, 2015.

21 ETSI. MEC in 5G networks. White Paper No. 28, First edition, June, 2018.

22 A. R. Prasad and S. Arumugam (eds). ICT Standardization. *Journal of ICT Standardization*, May 2018.

23 D. Soldani. 5G beyond radio access: A flatter sliced network. *Mondo Digitale*, AICA, March, 2018.

24 www.3gpagesorg.

25 3rd Generation Partnership Project. User Equipment (UE) radio transmission and reception; Part 1: Range 1 Standalone. 3GPP TS 38.101-1 V15.2.0.

26 3rd Generation Partnership Project. User Equipment (UE) radio transmission and reception; Part 3: Range 1 and Range 2 Interworking operation with other radios. 3GPP TS 38.101-3 V15.3.0.

27 Huawei. 5G Spectrum. Public policy position paper, March 2018.

28 3rd Generation Partnership Project. NR and NG-RAN overall description; Stage 2. 3GPP TS 38.300 v.15.2.0.

29 3rd Generation Partnership Project. System architecture for the 5G system; Stage 2. 3GPP TS 23.501 v.15.2.0.

30 3rd Generation Partnership Project. Procedures for the 5G System; Stage 2. 3GPP TS 23.502 v.15.2.0.

31 3rd Generation Partnership Project. NG-RAN; NG general aspects and principles. 3GPP TS 38.410 v.15.0.0.

32 3rd Generation Partnership Project. Security architecture and procedures for 5G System. 3GPP TS 33.501 v.15.1.0.

33 3GPP Security Aspects. See: http://www.3gpp.org/DynaReport/33-series.htm.

34 D. Soldani, M. Shore, J. Mitchell, and Mark Gregory. The 4G to 5G network architecture evolution in Australia. *The Australian Journal of Telecommunications and the Digital Economy (AJTDE)*, November, 2018.

35 D. Soldani, Y. J. Guo, B. Barani, P. Mogensen, C. L. I, and S. K. Das. 5G for Ultra-Reliable Low-Latency Communications. *IEEE Network, Special Issue on 5G for URLLC*, 32(2):6–7, 2018.

36 J. Sachs, G. Wikström, T. Dudda, R. Baldemair, and K. Kittichokechai. 5G radio network design for ultra-reliable low-latency communication. *IEEE Network, Special Issue on 5G for URLLC*, 32(2):24–31, 2018.

37 3rd Generation Partnership Project. NR; Physical channels and modulation. 3GPP TS 38.211 v.15.2.0.

38 G. Pocovi, H. Shariatmadari, G. Berardinelli, K. Pedersen, J. Steiner, Z. Li, and P. Marsch. Achieving Ultra-Reliable Low-Latency Communications: Challenges and envisioned system enhancements. *IEEE Network, Special Issue on 5G for URLLC*, 32(2):8–15, 2018.

39 J. Rao, S. Vrzic. Packet duplication for URLLC in 5G: Architectural enhancements and performance analysis. *IEEE Network, Special Issue on 5G for URLLC*, 32(2):32–40, 2018.

40 H.S. Park, Y. Lee, T.J. Kim, B. C. Kim, and J. Y. Lee. Handover mechanism in NR for ultra-reliable low-latency communications. *IEEE Network, Special Issue on 5G for URLLC*, 32(2):41–47, 2018.

41 I. Viering; H. Martikainen; A. Lobinger; B. Wegmann. Zero-zero mobility: Intra-frequency handovers with zero interruption and zero failures. *IEEE Network, Special Issue on 5G for URLLC*, 32(2):48–54, 2018.

42 Merrill Lynch, Bank of America. Robot Revolution – Global Robot & AI Primer. Equity, Nov. 2015.

43 Australian Centre for Robotic Vision. A robotics roadmap for Australia, 2018. See: https://www.roboticvision.org/robotic-computer-vision-roadmap/.

44 Financial Times. Artificial Intelligence and Robotics. FT top stories. See: https://www.ft.com/artificial-intelligence-robotics.

8

5G: Disruption in Media and Entertainment

Stamos Katsigiannis, Wasim Ahmad and Naeem Ramzan

Internet traffic is on the rise every year. Today, the total global Internet traffic is 1.6 zettabytes and Cisco is predicting that the traffic will double again by 2021. According to Cisco and Sandvine's latest report, video and music streaming make up more than 60% of global online data traffic, a number that will increase dramatically with the use of Augmented Reality (AR) and Virtual Reality (VR) in sports, gaming and other entertainment applications. This array of technologies will redefine the immersive experience of the users and will transform the way we spend our leisure time. Indeed, the advancement in technologies and the use of blended reality will influence both the entertainment industry and users greatly.

We are not far from when everything around us will be connected and interconnected, intelligent and communicative. When it comes to ensuring connectivity for a large number of devices and people, operators are all aware of the challenges they face. Some of these challenges include capacity, latency and traffic prioritization, to name a few. It is expected that 5G networks will address these issues and bring the breakthroughs into the real wireless world. 5G networks will not only empower high data speeds for streaming multimedia content on mobile devices, but will also reduce network latency drastically. It is predicted that transmission speeds with 5G will increase by at least tenfold when compared to 4G.

8.1 Multi-Channel Wireless Audio Systems for Live Production

Wireless Audio-Video (AV) equipment is widely used in concerts, theatre shows, musicals, and many other entertainment productions. The commonly used AV equipment in any live productions includes wireless cameras and microphones, in-ear monitors (IEMs), intercom devices, conference systems, and effect control. Only the live audio use case is discussed here. For live performances, it is highly desirable that the Professional Wireless Audio Devices (PWAD) provide highest quality-of-service and low-latency transmission. Therefore, appropriate communication technologies and interference-free radio resources are required to achieve these objectives.

Enabling 5G Communication Systems to Support Vertical Industries, First Edition.
Edited by Muhammad Ali Imran, Yusuf Abdulrahman Sambo and Qammer H. Abbasi.

In a typical live production setup, several wireless microphones are connected to the central mixing console, where each microphone signal is streamed through a dedicated channel to its corresponding receiver. The mixed audio from the central console is streamed back to the artist's IEM system on a different channel. Usually, the RF transmission is affected by human bodies and that effect surges because of the body-worn transmitters and receivers. In general, the RF performance is degraded because of the absorption and reflection by the surroundings as well as the frequency used – higher frequencies lead to higher degradation [1]. Therefore, most of the PWAD are currently placed below 2 GHz in order to scale down degradation.

Currently, audio equipment utilises the highly specialised and non-standardised TV UHF band that forms the core spectrum for PWAD. PWAD share the TV UHF band with broadcasting services, who are the primary users of this frequency range, but now this spectrum is becoming scarcer because part of its frequency chunk is going to be reassigned to the cellular industry [2]. Therefore, wireless audio device users are worried about the future of their UHF wireless systems, and are increasingly looking for new technology as a solution.

Professional Wireless Audio Systems (PWAS) are some of the most challenging applications, with spectral efficiency, latency, reliability, data rates, and synchronicity being of major importance. Some of these parameters are highly coupled and there is always a trade-off between them. The spectral efficiency in PWAS is the number of audio links transmitted per MHz in the employed spectrum [3]. Typically, PWAS use 200 kHz bandwidth per link and require a channel spacing between 400 and 600 kHz. Some PWASs offer precise numbers of 2.1 links/MHz on the required wireless transmission with 99.99 % reliability and approximately 3 ms end-to-end latency limit [4].

The latency in live performance is measured as the overall round-trip time from analog input at the microphone to analog output at the IEM. Professional musicians use monitor functionality on stage or during live production and demand that latency should not exceed 4 ms. By subtracting the processing time for Analog-to-Digital (A/D) conversion and the mixing console, a maximum latency of 1 ms per wireless audio link remains [4]. By comparing these figures with the existing PWAS, it becomes obvious that the latency target can only be achieved in combination with analog equipment.

Another strongly desirable feature for any live PWAS is reliability. Due to the live audio streaming and the strict latency requirement during live performance, it is not possible to repeat audio transmissions in the case of error. Therefore, robustness of the wireless communication service at the application level is the reliability measurement of audio packets that arrive without errors and within the defined time limit.

For the considered use case, the link data rate covers the range from 150 kb/s up to 4.61 Mb/s, while keeping the low latency and high reliability requirements. Normally, speech applications require a low data rate, but the highest data rates are utilised for the high definition productions where normally uncompressed audio is transmitted. Currently, typical data rates per link in live audio productions vary from 200 kb/s to 750 kb/s per device depending on the hardware used. Once the data rate is adjusted for one audio link, it remains constant during the whole operating time.

Synchronicity is another important feature required for wireless live professional production setups [4], which is described as the maximum allowed time offset between the reference signal and the device clocks. In the existing audio systems, the reference sampling interval is defined by the central clock and the sampling times in PWAS. An external clock generator, which is either a dedicated device or integrated generator, facilitates this central clock or word-clock.

5G can offer innovative approaches to solve the existing challenges in PWAS. Especially, the Ultra-Reliable Low-Latency Communications (URLLC) can offer useful technical solutions to achieve the desirable Key Performance Indicators (KPIs) of PWAS. The 5G standard also has the potential to replace TV UHF-based wireless technology in PWAD. The PWAS and URLLC use cases have strong resemblances and similar requirements. Therefore, there is a strong case to consider the possibilities of integrating the URLLC-based PWAS use case into the 5G ecosystem. Pilz et al. [5] suggested the potential integration of the PWAS in the upcoming frame structure of 5G New Radio (NR) for URLLC. By using this approach, the demand for additional capacity for the most demanding PWAS scenarios can also be covered by using already established techniques, e.g. Multiple-Input and Multiple-Output (MIMO) techniques. Based on the similarities and common features between the PWAS and other verticals, it is recommended to strengthen awareness of what is required to get such a professional application onboard.

8.2 Video

Recent years have seen the demand for video streaming over mobile networks growing exponentially. The continuously increasing quantity of available video content, either in an on-demand form or broadcasted live, and the requirement for higher resolutions are pushing the limits of current 4G networks. Forecasts show that video-over-mobile traffic will represent over three fourths (78%) of all Internet traffic by 2021 [6] and as a result, 4G-LTE technology will not be sustainable in the long term. 5G can provide a solution to this issue by providing significantly higher data bandwidth and low-latency communications that will not only accommodate the increased demand for high-resolution video over mobile networks but also allow for the consumption of more immersive video modalities such as 360° video.

8.2.1 Video Compression Algorithms

Sophisticated video compression algorithms are required in order to take advantage of the increased bandwidth available in modern 5G networks. The following state-of-the-art video compression algorithms are able to support high-resolution video streams at low bit rates and will be the core components of 5G video-over-mobile services.

8.2.1.1 HEVC: High Efficiency Video Coding

Known also as H.265, High Efficiency Video Coding (HEVC) [7] is the latest video compression standard by the Joint Collaborative Team on Video Coding (JCT-VC), consisting of the International Organization for Standardization and the International Electrotechnical Commission's Moving Picture Experts Group (ISO/IEC MPEG) and the International Telecommunication Union's Telecommunication Standardization Sector Video Coding Experts Group (ITU-T VCEG). It was designed as a successor to the widely used AVC/H.264 [8] compression standard, with a target of up to 50%

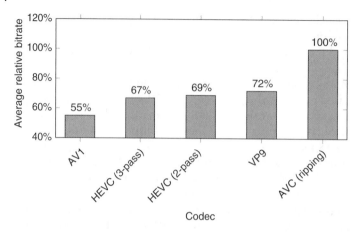

Figure 8.1 Average relative bit rate for a fixed quality. The x265 and x264 open-source encoder libraries were used for HEVC/H.265 and AVC/H.264 respectively. The reference encoders were used for AV1 and VP9.

reduction in bit rate for the same resolutions or significant increase in visual quality for the same sizes, in comparison to AVC/H.264. As shown in Figure 8.1, recent comparative studies show that HEVC is able to produce the same quality as H.264 at 67% of the data rate [9]. HEVC can support resolutions of up to 8192 × 4320 (4320p 8K Ultra High Definition – UHD) and all the modern high-resolution standards. The first version of the HEVC standard was ratified in 2013. Subsequent versions published in 2015 added extensions for multiview (MV-HEVC), range (RExt), and scalability (SHVC), while extensions for 3D video (3D-HEVC) and Screen Content Coding (SCC) were added in early 2017, allowing the use of rendered graphics, text, and animations along with regular video scenes [10, 11]. The use of highly optimised implementations of the HEVC standard in combination with 5G mobile networks has the potential to accommodate the need for high-resolution and high-quality video-over-mobile network services, for both the video-on-demand and live broadcasting/streaming scenarios.

8.2.1.2 VP9

VP9 [12] is an open-source royalty-free video compression format developed by Google Inc. which was finalised in 2013. It is an evolution of Google's VP8 compression algorithm, aiming to achieve a 50% reduction in bit rate. As a main competitor to HEVC, which charges patent royalties to its commercial users, the royalty-free nature of VP9 constitutes one of its major advantages. Its available coding tools include multiple prediction block sizes, multiple intra-predictors, sub-pixel interpolation, three different transform types, and entropy encoding. The maximum supported resolution is 8192 × 4352 at up to 120 fps, allowing for 8K UHD resolutions. Furthermore, VP9 supports various colour spaces, such as sRGB, SMPTE-170, SMPTE-240, Rec 601, and others. Initial comparisons with HEVC were in favour of the latter [13, 14]. Nevertheless, recent comparative studies from Moscow State University (MSU) showed that HEVC produced the same quality as H.264 at 67% of the data rate, with VP9 close behind at 72% [9], as shown in Figure 8.1. As of 2018, the most widespread use of the VP9 compression algorithm is on Google's popular video platform YouTube [15].

8.2.1.3 AV1: AOMedia Video 1

In an attempt to provide a royalty-free next generation video compression algorithm, the Alliance for Open Media (AOMedia) initiated the development of the AOMedia Video 1 (AV1) specification [16]. The AOMedia consortium includes established companies and organizations in the field of software, video-on-demand, semiconductors, and web browsers, such as Google, Microsoft, Netflix, Mozilla, Cisco, and Amazon among others. AV1 is based on Google's VP9 specification [12] and incorporates contributions from Google's VP10, Xiph's/Mozilla's Daala, and Cisco's Thor video compression algorithms. The AV1 bitstream specification was released on March 2018, with version 1.0.0 of the specification following on June 2018 [16]. AV1 follows the typical hybrid block-based approach used in previous MPEG/ITU standards from the MPEG/ITU family and aimed to achieve a 25% improvement over HEVC. Recent studies showed similar performance between AV1 and HEVC [17], while the comparative study from MSU showed that AV1 produced the same quality as H.264 at 55% of the data rate, with HEVC close behind at 67% [9], as shown in Figure 8.1.

8.2.2 Streaming Protocols

Streaming video over a network requires more than just a sophisticated video compression algorithm. The multitude of user requirements and capabilities in terms of available bandwidth and playback equipment necessitate the use of adaptive and scalable protocols for video streaming. The basic approach of such protocols is the creation of multiple versions of the same video, typically using different bit rates or spatial resolutions, and the division of these versions into small segments (typically few seconds). The video segments can then be provided by a web server and accessed through standard HyperText Transfer Protocol (HTTP) GET requests. A manifest is provided to the client prior to the streaming session, describing the available qualities of the segment and information such as start time, duration, etc., along with their unique HTTP Uniform Resource Locators (URLs). Consequently, before video streaming commences, a client will retrieve the manifest and request the video segments that best fit its requirements. Adaptation to different bit rates and spatial resolutions is handled by the client on a per-segment basis by requesting the appropriate video segments. As a result, mobile users can move through physical locations while maintaining a smooth video streaming experience. For example, a user can receive a high-resolution video stream while using his/her fast home wireless internet connection, then automatically switch to a medium resolution stream after walking out the door where high-speed mobile connectivity is available, and even switch to a low-resolution stream in areas where mobile service is poor. Among the various video streaming protocols that exist, Apple HLS [18] and MPEG DASH [19] are the most dominant.

8.2.2.1 Apple HTTP Live Streaming (HLS)

Apple HTTP Live Streaming (HLS) [18] is a protocol for video streaming over the Internet, originally developed by Apple. HLS delivers video content via standard HTTP web servers and is widely supported by various hardware and software platforms. The use of standard HTTP web servers make the HLS protocol compatible with standard Content Distribution Networks (CDNs) and less prone to blocking by firewalls and network traffic restrictions. Videos are typically cut into segments of 10 seconds and can be encoded with either AVC/H.264 or HEVC/H.265. Furthermore, bit rate and quality adaptation can be performed by both the client and the server by dynamically detecting the network speed and adjusting video quality accordingly. HLS also supports other features such as embedded captions, DRM, multiple-stream synchronised playback, various advertising standards, and lower latency streaming.

8.2.2.2 Dynamic Adaptive Streaming over HTTP (DASH)

Dynamic Adaptive Streaming over HTTP (MPEG-DASH) [19] is an adaptive bit rate HTTP-based streaming protocol that has been standardised by MPEG as ISO/IEC 23009-1 in April 2012. Similarly to HLS, MPEG-DASH works by dividing the videos into smaller segments in a variety of bit rate and quality settings and delivering them to clients via standard HTTP web servers. Bit rate and quality adaptation is performed at the client by automatically selecting the most suitable next segment for download and playback, according to current network conditions. By always selecting the segments with the highest bit rate that can be downloaded without leading to re-buffering and playback interruption, MPEG-DASH achieves smoother playback under changing network conditions. Furthermore, unlike HLS which only allows AVC/H.264 and HEVC/H.265 encoded video segments, MPEG-DASH is codec-agnostic, thus permitting any available video coding algorithms and providing increased flexibility to video content providers. Additional features of MPEG-DASH include support for DRM, advertising platforms, subtitles, lower latency streaming, and others.

8.2.3 Video Streaming Over Mobile Networks

Table 8.1 provides information about the bandwidth requirements for streaming pre-encoded video when using the AVC/H.264 video compression algorithm which is the current industry and market standard. Live video streaming generally requires higher bit rates compared to pre-encoded videos as a result of sub-optimal compression due to the real-time requirements. Taking into consideration the reported average download speeds for mobile networks in the world (Table 8.1), it is evident that current 4G-LTE mobile networks are not able to reliably support video streaming at a higher resolution than 1440p (2K) @30fps, while 3G networks struggle even with High Definition (HD – 720p) and Full High Definition (FullHD – 1080p) video streaming, depending on the quality of mobile service and the location. Maximum theoretical download speeds support the streaming of higher resolution content via 3G and 4G-LTE networks, but such speeds are rarely achieved in practice and vary significantly across different countries and across different regions within a country. The use of more state-of-the-art video compression algorithms, such as HEVC/H.265 and AV1, which can achieve up to 50% reduction in the required bit rate, will enable video streaming of higher resolutions via 3G and 4G-LTE networks, albeit being limited to 2K resolutions for pre-encoded video in practice, and to even lower resolutions for live streamed video. Considering the capabilities of modern video cameras, as well as the proliferation of virtual reality and 360° video equipment that requires resolutions of 4K or more, it is evident that current 4G-LTE and older mobile networks are not

Table 8.1 Video streaming capabilities of mobile networks in practice and in theory

Video type	Bit rate H.264 (Mbps)	2G (0.1/0.3)[a] (Mbps)	3G (4.1/42)[a] (Mbps)	4G (16.9/100)[a] (Mbps)	5G (1-10)[b] (Gbps)
4320p @60fps (8K)[c]	100-120	× /×	× /×	× /×	✓
4320p @30fps (8K)[c]	70-90	× /×	× /×	× /✓	✓
2160p @60fps (4K)[d]	53-68	× /×	× /×	× /✓	✓
2160p @30fps (4K)[d]	35-45	× /×	× /✓	× /✓	✓
1440p @60fps (2K)[d]	24	× /×	× /✓	× /✓	✓
1440p @30fps (2K)[d]	16	× /×	× /✓	✓ /✓	✓
1080p @60fps (HD)[d]	12	× /×	× /✓	✓ /✓	✓
1080p @30fps (HD)[d]	8	× /×	× /✓	✓ /✓	✓
720p @60fps (HD)[d]	7.5	× /×	× /✓	✓ /✓	✓
720p @30fps (HD)[d]	5	× /×	× /✓	✓ /✓	✓
480p @60fps (SD)[d]	4	× /×	✓ /✓	✓ /✓	✓
480p @30fps (SD)[d]	2.5	× /×	✓ /✓	✓ /✓	✓
360p @60fps (SD)[d]	1.5	× /×	✓ /✓	✓ /✓	✓
360p @30fps (SD)[d]	1	× /×	✓ /✓	✓ /✓	✓

a) Average reported download connection speed in the world as of February 2018. *Source: OpenSignal – The State of LTE (February 2018)*/Theoretical maximum download speed for 2G-EDGE, 3G-DC-HSPA+, 4G-LTE.
b) Theoretical expected download speed.
c) Recommended upload encoding settings for SproutVideo (November 2018).
d) Recommended upload encoding settings for YouTube (November 2018).

able to support the increased demand for bandwidth. On the other hand, with expected theoretical download speeds between 1 and 10 Gbps and a latency of approximately 1 *ms*, 5G networks are the ideal platform for streaming high-resolution video content and can spearhead the mass adoption of next generation media formats, including 4K, 8K, and 360° video, virtual reality applications, etc.

8.3 Immersive Media

The increased bandwidth capacity and decreased latency of 5G networks will allow access to complex immersive audio-visual experiences over mobile networks. The recent availability of commercial-grade Head-Mounted Displays (HMDs) and support from mobile phone/tablet devices has created a developing market and a massive consumer audience for VR, AR, and 360° video content over mobile networks. With studies projecting that the VR/AR market would grow to $2.16 trillion by 2035 [20] as more and more industries adopt VR/AR technology [21], it is evident that 5G networks would be instrumental in this growth.

8.3.1 Virtual Reality (VR)

VR refers to a 3D computer-generated environment that users can explore and interact with, using specialised equipment. VR environments typically offer a 360° view capability and may either resemble actual physical environments or be completely artificial, thus providing enhanced flexibility to content creators and allowing the users to engage in novel interactive experiences. Head-mounted displays (Figure 8.2) are the most common equipment for accessing VR content, due to low cost, availability of equipment, and the immersion levels that they can offer. Specially designed rooms equipped with multiple large screens can also be used but at a much larger cost and in a non-portable manner. Smaller devices that have the ability to present three-dimensional images, such as smart-phones, tablets, and computers, may also be used, albeit offering a less immersive experience. Additionally, while VR is usually focused on auditory and visual feedback, it may incorporate various other types of sensory feedback, such as haptic or smell. There is a variety of devices and sensors that can contribute to the immersion of the user into the virtual environment, including force-feedback sensors, seats and other motion platforms, omnidirectional treadmills, nasal wearables for smell creation, and others. Employing such a diverse set of sensory input allows the user to become more immersed into the virtual environment and offers a more realistic experience. Exploiting the immersion that it offers, VR technology is being increasingly used in fields like entertainment, video games, training, military simulations, construction, architecture, design, and others. Some example applications of VR are provided in Table 8.2.

8.3.2 Augmented Reality (AR)

AR, also known as mixed reality or computer-mediated reality, entails the creation of an enhanced version of reality by superimposing computer-generated content over the user's view of the real world, thus augmenting the user's perception of reality. The superimposed computer-generated content is not limited to images and video but can be expanded across multiple sensory modalities (e.g. visual, auditory, haptic, etc.) in

Figure 8.2 Head-mounted display for 360°/VR/AR applications. *Published under the CC0 Creative Commons licence.*

Table 8.2 Example applications of virtual and augmented reality.

Field	Application	VR	AR
Archaeology	Superimposing archaeological features onto modern landscapes		✓
Archaeology	Recreation of archaeological sites	✓	✓
Architecture	Visualization of building projects	✓	✓
Commerce	Superimposing marketing material on the real world		✓
Education	Superimposing multimedia content and other information on a blackboard		✓
Engineering	Visualization of new designs	✓	✓
Live events	Superimposing interesting information and commentary		✓
Live events	Roaming though the environment and selecting the viewpoint	✓	
Medicine	Enhancing telemedicine capabilities	✓	✓
Military	Information overlay on maps and visual equipment		✓
Navigation	Display of information on vehicle windscreens		✓
Tourism	Objects of tourist interest rendered on the landscape		✓
Video games	Digital gaming within a virtual environment	✓	
Video games	Digital gameplay in a real world environment		✓
Workplace	Facilitate collaboration among distributed team members	✓	✓

order to create rich interactive experiences of a real-world environment, either live or recorded. The addition of artificial components on the real-world view can either be constructive, i.e. adding information to the real world, or destructive, i.e. masking (removing) information from the real world and replacing it with artificial information. The most important advantage of AR is that it does not change the user's perception of the real world by just displaying data, but through providing immersive sensory stimuli that is perceived as an actual part of the world.

Early AR applications focused mostly on the entertainment and gaming industries but current advances in the field have led to major interest from sectors like education, professional training, knowledge sharing, collaborative environments, and others. In addition, the release of Apple's ARKit and Google's ARCore in 2017 gave developers powerful frameworks for creating AR applications for the most widespread mobile platforms (iOS and Android), thus increasing the potential market penetration of AR technology. A list of potential AR applications can be found in Table 8.2.

8.3.3 360-Degree Video

360° videos are video recordings that capture a view towards every direction at the same time. Omnidirectional cameras or multiple regular cameras are used for recording such videos. In the latter case, overlapping angles of the video are filmed simultaneously and video stitching and colour and contrast calibration methods are employed in order to merge the footage into one spherical video piece. Normal flat displays, such as the ones found in computers, TV monitors, smart-phones and tablets can be used for playback, by displaying only the video region within the field-of-view of the viewer. At the same time, the viewers are able to control the viewing direction similarly to panorama imagery. 360° video can also be displayed using multiple projectors projecting on spherical surfaces for a more cinematic experience or using HMDs for more portability and comfort. An equirectangular projection is typically used in order to encode 360° video using generic video compression algorithms and the encoded video can either be monoscopic or stereoscopic, i.e. displaying the same image to both eyes or having a distinct image directed at each eye respectively. 360° video technology expands video viewing to a more immersive experience that allows the audience to watch a video from multiple viewing angles and not be constrained by the viewing experience that the video creators originally envisioned. Furthermore, it can potentially revolutionise the way that remote viewers experience live events like sports matches, music concerts, sightseeing, and others. By using multiple omnidirectional cameras and broadcasting 360° video of the events, the viewers will have the ability to virtually roam through the venue of the events and select their favourite viewpoint to watch the event, join the acts, or stand next to the performers or athletes.

8.3.4   Immersive Media Streaming

The most simple and straightforward approach for streaming 360°/VR content is to stream a video that constitutes an equirectangular projection of the 360° panorama. Common state-of-the-art video compression algorithms, such as AVC/H.264, HEVC/H.265, VP9, etc., can then be used to compress the video and video streaming protocols such as HLS or MPEG-DASH are employed for adaptive HTTP streaming. The decoding of the video and recreation of the 360° panorama is handled at the receiving client. The simplicity of this approach comes at an immense cost in bandwidth, as the regions of the video that are not inside the field-of-view have to be continuously transmitted and decoded. Using this approach to stream 4K 360°/VR requires more than 20 Mbit/s using modern video compression approaches, while 8K 360°/VR requires more than 80 Mbit/s [22]. It is evident that such content streaming pushes the limits of current 4G mobile networks and requires a massive increase in available bandwidth in order for the technology to take off. 5G networks can be the solution to this issue and provide the necessary mobile network speeds to allow for the mass consumption of 360°/VR content.

Apart from increasing network speeds, there have been various attempts to reduce the bandwidth requirements for 360°/VR streaming, most of which focus on selective transmission and rendering of only the field-of-view and the neighbouring regions of the video [23]. Such approaches require the separate encoding of video regions which can then be transmitted in a way similar to the video segments in adaptive HTTP video streaming approaches. The challenge in such approaches lies in switching fast enough to the suitable video segment when a user moves the field-of-view. Other approaches, such as "Tiled VR streaming" [22], rely on streaming a low-resolution version of the whole 360° panorama and only stream high-resolution tiles (video regions) relevant to the field-of-view of the user. Such approaches can achieve smoother transitions between the fields-of-view and reduce the effects of simulator sickness symptoms by accomplishing an instant response to moving, due to the low-resolution base layer. Nevertheless, they require latency in the range of 20-40 *ms* [22] in order to stream the high-resolution tiles on time. Both approaches for reducing the bandwidth requirements for high-resolution 360°/VR streaming require the low latencies that 5G networks can offer.

5G mobile networks have the ability to provide both the increased bandwidth required for straightforward high-resolution streaming of 360°/ VR panorama videos, as well as the low latency needed to achieve smooth 360°/VR rendering over mobile networks when lower bandwidth approaches are utilised. The proliferation of 5G networks will spearhead the adoption of 360°/VR technology in various sectors of the industry and everyday life and become the catalyst for the expansion of the relevant market. Brief examples of how 360°/VR technology can benefit various fields are provided in Table 8.2.

References

1 E. Cabot and M. Capstick. *The effect of the human body on wireless microphone transmission.* ITIS Foundation, Final Report Project 516, July 2015.

2 Bundesnetzagentur. Strategic aspects of the availability of spectrum for broadband rollout in Germany, January 2017.

3 European Telecommunications Standards Institute. Audio PMSE up to 3 GHz, Part 1–3: Class [A,B,C] Receivers. ETSI EN 300 422-1, V2.1.2, January 2017.

4 PMSE-XG Project. PMSE and 5G. White Paper, March 2017. URL http://pmse-xg .research-project.de/. [Online; accessed 15 November 2018].

5 J. Pilz, B. Holfeld, A. Schmidt, and K. Septinus. Professional live audio production: A highly synchronized use case for 5G URLLC systems. *IEEE Network*, 32(2):85–91, March 2018. ISSN 0890-8044. doi: 10.1109/MNET.2018.1700230.

6 Cisco Systems Inc. Cisco visual networking index: Global mobile data traffic forecast update, 2016–2021, February 2017. (White paper). Document ID:1454457600805266.

7 ITU-T Rec. H.265. H.265: High efficiency video coding, December 2016.

8 ITU-T Rec. H.264. H.264: Advanced video coding for generic audiovisual services, April 2017.

9 Moscow State University. MSU codec comparison 2017 Part V: High quality encoders. Technical report, Moscow State University, 2017.

10 G. J. Sullivan, J. Ohm, W. Han, and T. Wiegand. Overview of the high efficiency video coding (hevc) standard. *IEEE Transactions on Circuits and Systems for Video Technology*, 22(12):1649–1668, Dec 2012. ISSN 1051-8215. doi: 10.1109/TCSVT.2012.2221191.

11 T. K. Tan, R. Weerakkody, M. Mrak, N. Ramzan, V. Baroncini, J. Ohm, and G. J. Sullivan. Video quality evaluation methodology and verification testing of hevc compression performance. *IEEE Transactions on Circuits and Systems for Video Technology*, 26(1):76–90, Jan 2016. ISSN 1051-8215. doi: 10.1109/TCSVT.2015.2477916.

12 WebM. VP9 bitstream & decoding process specification. WebM VP9 specification Version 0.6, 2017. URL https://www.webmproject.org/vp9/. [Online; accessed 8 November 2018].

13 D. Grois, D. Marpe, T. Nguyen, and O. Hadar. Comparative assessment of H.265/MPEG-HEVC, VP9, and H.264/MPEG-AVC encoders for low-delay video applications. In *Proc. SPIE 9217, Applications of Digital Image Processing XXXVII*, 2014. doi: 10.1117/12.2073323.

14 M. Řeřábek and T. Ebrahimi. Comparison of compression efficiency between HEVC/H.265 and VP9 based on subjective assessments. In *Proc.SPIE*, volume 9217, 13 pages, 2014. doi: 10.1117/12.2065561.

15 S. Robertson. VP9: Faster, better, buffer-free youtube videos, April 2015. URL https://youtube-eng.googleblog.com/2015/04/vp9-faster-better-buffer-free-youtube.html. [Online; posted 6 April 2015].

16 AOMedia. AV1 bitstream & decoding process specification. The Alliance for Open Media, November 2018. URL https://aomediacodec.github.io/av1-spec/av1-spec.pdf. [Online: accessed 15 November 2018].

17 A. S. Dias, S. Blasi, F. Rivera, E. Izquierdo, and M. Mrak. An overview of recent video coding developments in MPEG and AV1. IBC, September 2018. URL https://www.ibc.org/content-management/an-overview-of-recent-video-coding-developments-in-mpeg-and-av1/3303.article.

18 RFC 8216. HTTP live streaming. Apple Inc., August 2017.

19 ISO/IEC 23009-1. Dynamic adaptive streaming over HTTP (DASH) – Part 1: Media presentation description and segment formats, December 2017.

20 K. Ezawa, C. Danely, J. B. Bazinet, A. Shirvaikar, W. H. Pritchard, D. Chan, M. May, T. Kanai, J. Suva, and T. Fujiwara. Virtual and augmented reality: Are you sure it isn't real? Citi Research, December 2015. URL https://www.citivelocity.com/rendition/eppublic/uiservices/open_watermark_pdf?req_dt=cGRmTGluaz1odHRwcyUzQSUyRiUyRmlyLmNpdGkuY29tJTJGdXha QlZoRHclMjUyRkFIM1JseURGMHhNVWRNUGJBSWpiWmVVbWJuTDN0 ZXhESSUyNTJCa0lzeTI5QTFGNUElMjUzRCUyNTNEJnVzZXJfaWQ9amIy MTE1NSZ1c2VyX3R5cGU9RQ. [Online: accessed 15 November 2018].

21 G. Munster, T. Jakel, D. Clinton, and E. Murphy. Next mega tech theme is virtual reality. Piper Jaffray Investment Research, May 2015. URL http://cdn.instantmagazine.com/upload/4666/piperjaffray.f032beb9cb15.pdf. [Online; accessed 15 November 2018].

22 R. van Brandenburg, R. Koenen, and D. Sztykman. CDN optimization for VR streaming. IBC, October 2017. URL https://www.ibc.org/tech-advances/cdn-optimisation-for-vr-streaming-/2457.article?adredir=1.

23 E. Kuzyakov and D. Pio. Next-generation video encoding techniques for 360 video and VR. Facebook Code, January 2016. URL https://code.fb.com/virtual-reality/next-generation-videoencoding-techniques-for-360-video-and-vr/. [Online; accessed 15 November 2018].

9

Towards Realistic Modelling of Drone-based Cellular Network Coverage

Haneya Naeem Qureshi and Ali Imran

The demand for more diverse, flexible, accessible and resilient broadband service with higher capacity and coverage is on the rise. Some of these requirements can be accomplished with unmanned aerial vehicles (UAVs) acting as aerial base stations. This is because of the several advantages UAV-based communication offers, such as higher likelihood of line-of-sight (LoS) path and less scatter and signal absorption as compared to terrestrial systems [1, 2]. Moreover, the demand for increase in capacity is leading towards deployment of small cells in terrestrial networks, resulting in the need for higher cell counts, leading to far larger number of ground sites. This makes the goal of attaining seamless coverage over a wide geographical area through terrestrial systems unfeasible due to limited availability of suitable sites and local regulations. This challenge is likely to aggravate with advent of even smaller mmWave cells offering even more sporadic coverage [3].

Similarly, satellite networks have their own limitations such as high latency, high propagation loss, limited orbit space and high launching costs. On the contrary, UAVs can be deployed quickly with much more flexibility to move from one point to another, which is a desirable feature for rapid, on-demand or emergency communications [4].

UAVs can thus be seen as potential enablers to meet the several challenges of next generation wireless systems by either functioning as complementary architecture with already existing cellular networks to compensate for cell overload during peak times and emergency situations [5, 6] or by serving as stand-alone architecture to provide new infrastructure, especially in remote areas [7, 8].

However, in order to fully reap the benefits of UAV-based communication, optimal design of UAVs deployment parameters is of fundamental importance. In this chapter, we address the UAV deployment problem by analyzing the trade-offs between key system design parameters such as height, antenna beamwidth, and number of UAVs. By leveraging a more realistic model compared to prior studies on the topic, the analysis reveals several new insights and trade-offs between the design parameters that remain unexplored in existing studies.

Enabling 5G Communication Systems to Support Vertical Industries, First Edition.
Edited by Muhammad Ali Imran, Yusuf Abdulrahman Sambo and Qammer H. Abbasi.
© 2019 John Wiley & Sons Ltd. Published 2019 by John Wiley & Sons Ltd.

9.1 Overview of Existing Models for Drone-Based Cellular Network Coverage

Several studies have recently addressed UAV deployment for different service requirements, mostly using altitude, transmission power and number of UAVs as the only three deployment parameters. For example, the maximum coverage and optimal altitude assuming one UAV with no interference can be estimated as a function of maximum allowed path loss and statistical parameters of urban environment [9]. However, this approach is limited to a single UAV while using mean value of shadowing (rather than its random behaviour) and altitude as the only optimization parameter to control coverage. Another way to determine optimal height for maximum coverage for a single UAV is based on coverage probability and the information rate of users on the ground at a particular UAV altitude [10]. Optimal altitude for both maximum coverage and minimum required transmit power for two UAVs can be found in [11].

Changes in coverage area also affects other performance indicators, such as carrier to interference ratio [12] and Receive Signal Strength Indicator (RSSI). Analysis using some of popular models, such as Okumura-Hata, COST-Hata, and COST Walfish-Ikegami (COST-WI) shows that signal strengths decrease faster with increase in altitude [13]. However, such models can only be used to deploy UAVs up to an altitude of 500m because of the underlying path loss models constraints.

Other deployment models consist of finding the optimal altitude of UAV based on the desired radius of coverage in real time [14] and numerically by considering the altitude of UAV and the location of both UAV and users in the horizontal dimension [15]. Ideas introduced in [15] are further elaborated in [16], leading to the conclusion that larger buildings require a higher UAV altitude. Additionally, gain adjustment strategy for circular beam antennas can be leveraged for optimal deployment by considering the impact of antenna power roll-off in 3G networks at a fixed platform height [17].

However, none of the aforementioned studies consider the impact of antenna gain pattern on the coverage versus height trade-off. One recent study that takes into account the effect of directional antenna considers joint altitude and beamwidth optimization for UAV-enabled multiuser communications [18]. In [18], the authors consider three communication models: downlink multicasting, downlink broadcasting and uplink multiple access, and conclude that optimal beamwidth and height critically depend on the communication model considered. However, this study uses line-of-sight propagation conditions and a step-wise antenna gain model for analytical tractability. Another recent study [19] that does consider the effect of antenna also uses a step-wise antenna gain model with only two possible values of antenna gain. This chapter aims to fill this gap by studying the optimization of UAV deployment design parameters while using a realistic 3D directional antenna model in the system. The analysis presented in this chapter shows that the use of a realistic antenna pattern makes a trend shifting difference in the height versus coverage trade-off and adds a new dimension of beamwidth to the UAV deployment design space.

9.2   Key Objectives and Organization of this Chapter

The organization and key objectives of this chapter are summarized as follows:

- Development of a mathematical framework for UAV deployment design while incorporating a realistic model for a practical directional antenna. Current studies on UAV deployment either ignore the effect of 3D directional antenna [9, 10, 15, 16, 20] or consider an over-simplified model for antenna gain [18, 19]. This gap is addressed by using a 3GPP-defined 3D parabolic antenna pattern whose gain is realistically dependent not only on beamwidth but also on three-dimensional elevation angle (Section 9.4 and Section 9.5.1).
- Derivation of analytical expressions for coverage characterized by received signal strength (RSS) as a function of height, beamwidth and coverage radius (Section 9.5.2).
- Development of an analytical framework to quantitatively analyze trade-offs among cell radius, beamwidth and UAV height. Section 9.6 investigates these trade-offs in light of the interesting interplay among the five key factors that define UAV-based coverage design space: antenna beamwidth and angular distance dependent gain, elevation angle dependent probability of line of sight, shadowing, free space path loss and height.
- Analysis of the impact of key UAV design parameters on RSS and validation of the derived expressions for probability density function (PDF) of RSS through simulations (Section 9.7).
- Extension of the proposed framework to a range of frequencies and environments (Section 9.8).
- Investigation of factors that can be used to control coverage of UAV. Contrary to the findings from prior studies [9, 10, 19] which use UAV altitude as the only optimization parameter to control coverage, based on the joint analysis of effect of beamwidth and altitude on coverage, it is shown that:
  – There exists an optimal beamwidth for given height for maximum coverage radius and vice versa.
  – Antenna beamwidth is a more practical design parameter to control coverage instead of UAV altitude. This is done by performing comparative analysis of the two by quantifying the sensitivity of coverage to both height and beamwidth (Section 9.9).
- Analysis of coverage probability pattern with varying tilt angles and asymmetrical beamwidths (Section 9.10), which highlights the capability of the derived equations and the underlying system model to extend the analysis to a wide range of scenarios, such as non-zero tilt angle and asymmetrical beamwidths.
- Extension of the analysis to multiple UAVs (Section 9.11). Some recent studies have leveraged circle packing theory to determine the number of UAVs needed to cover a given area [20]. However, this approach has two caveats: 1) it leaves significant coverage holes when two or more UAVs are used to cover an area and 2) the number of

UAVs increases dramatically with the increase in required coverage probability. To circumvent the problems posed by circle packing theory, the use of hexagonal packing is proposed and the results are compared with those obtained by circle packing. This comparison identifies several further advantages of proposed approach.

- Conclusion of the chapter (Section 9.12).

9.3 Motivation

Current studies on UAV-based cellular coverage adopt an oversimplified antenna pattern, which can lead to significant deviation from the performance trends with actual antenna patterns. Essentially, the flat-top antenna pattern quantizes the continuously varying antenna array gains in a binary manner. Although such a model significantly simplifies the analytical derivation, the oversimplified flat-top pattern yields results on optimal height, coverage radius and number of UAVs that may not hold for real UAV deployments, as will be revealed in this chapter. Moreover, it is difficult to analyze the impact of directional antenna with the flat-top antenna pattern, as only a few parameters are extracted to abstractly depict the actual antenna pattern. Particularly, the effect of some critical parameters of the antenna beam pattern such as horizontal and vertical beamwidths, tilt angle, azimuth angle, and elevation angle on coverage probability remain hidden in the binary-gain antenna model. Moreover, the binary quantization of the array gains cannot reflect the roll-off characteristic of the actual antenna pattern and therefore is unable to provide different gains for various azimuth and elevation angles.

To this end, an antenna pattern that approximates the actual antenna pattern accurately and realistically is required to reveal more insights in order to properly dimension and optimize UAV-based networks. In this chapter, this problem is addressed by using a 3GPP defined 3D parabolic antenna pattern whose gain is realistically dependent on not only beamwidth but also the three-dimensional elevation angle.

By finding new analytical results of coverage probabilities that capture the relative effect of all factors that impact the coverage radius concurrently – angular distance-dependent realistic non-linear antenna model, elevation angle-dependent probability of line of sight, shadowing, and a measurement-backed path loss model – several new insights into UAV coverage trade-offs are obtained. The analysis presented in this chapter shows that the use of a realistic antenna pattern makes a trend shifting difference in designing and optimizing UAV-based networks and also adds a new dimension of beamwidth to the UAV deployment design space in different scenarios.

9.4 System Model

We consider a system model illustrated in Figure 9.1. The UAV resides at a height h and projects a cell with coverage radius r. We define UAV coverage area as a set of points in circle of radius r, where a mobile station (MS) experiences a RSS, S_r above a threshold, γ. Here r is measured from the projection of UAV on ground when tilt angle, $\phi_{tilt} = 0$. In Figure 9.1, ϕ_{tilt} is the tilt angle in degrees of the antenna mounted on UAV, ϕ_{MS} is the vertical angle in degrees from the reference axis (for tilt) to the MS. θ_a is the angle

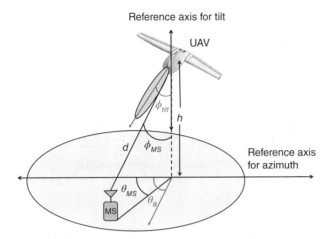

Figure 9.1 System model.

of orientation of the antenna with respect to horizontal reference axis, i.e. positive x-axis, and θ_{MS} is the angular distance of MS from the horizontal reference axis.

Utilizing the geometry in Figure 9.1, the perceived antenna gain from the UAV using a three-dimensional antenna model recommended by 3GPP [21] at the location of MS can be represented as in (9.1). Here B_{EW} and B_{NS} represent the beamwidths of the UAV antenna in East-West and North-South directions respectively, while λ_{EW} and λ_{NS} represent the weighting factors for the beam pattern in both directions respectively. G_{max} and A_{max} denote the maximum antenna gain in dB at the boresight of the antenna and maximum attenuation at the sides and back of boresight respectively.

$$G(\phi_{MS}, B_{NS}, B_{EW}) = \lambda_{NS} \left(G_{max} - \min\left(12\left(\frac{\phi_{MS} - \phi_{tilt}}{B_{NS}} \right)^2, A_{max} \right) \right)$$

$$+ \lambda_{EW} \left(G_{max} - \min\left(12\left(\frac{\theta_{MS} - \theta_a}{B_{EW}} \right)^2, A_{max} \right) \right) \tag{9.1}$$

The air-to-ground channel can be characterized in terms of probabilities of LoS and non-line-of-sight (NLoS) links between the UAV and MS. The probability of LoS link is estimated as [22]:

$$P_l(\phi_{MS}) = 0.01j - \frac{0.01(j - k)}{1 + \left(\frac{90 - \phi_{MS} - l}{m} \right)^n} \tag{9.2}$$

where $(j, \ldots n)$ are the set of empirical parameters for different types of environments and are given in Table 9.1(a). The angle $90 - \phi_{MS} = \tan^{-1}(\frac{h}{r})$ is the angle of elevation of the MS to the UAV. The probability of NLoS link is then $1 - P_l(\phi_{MS})$.

In addition to free space path loss, UAV-MS signal faces an elevation angle dependent shadowing. The mean and standard deviation for this shadowing can be modelled as [22]:

$$\mu_{sh} = \frac{p_\mu + (90 - \phi_{MS})}{q_\mu + t_\mu(90 - \phi_{MS})}, \qquad \sigma_{sh} = \frac{p_\sigma + (90 - \phi_{MS})}{q_\sigma + t_\sigma(90 - \phi_{MS})} \tag{9.3}$$

Table 9.1 Environment and frequency dependent parameters.

	Suburban	Highrise urban
j	101.6	352.0
k	0	−1.37
l	0	−55
m	3.25	173.80
n	1.241	4.670

f (GHz)		p_y	q_y	t_y
2.0	$v = \mu$	−94.20	−3.44	0.0318
	$v = \sigma$	−89.55	−8.87	0.0927
3.5	$v = \mu$	−92.90	−3.14	0.0302
	$v = \sigma$	−89.06	−8.63	0.0921
5.5	$v = \mu$	−92.80	−2.96	0.0285
	$v = \sigma$	−89.54	−8.47	0.9000

(a) Environment dependent parameters for P_l.　　　(b) Frequency dependent parameters for shadowing.

where p_μ, q_μ, t_μ, p_σ, q_σ and t_σ are parameters obtained from empirical measurements given in Table 9.1(b). In Table 9.1(b), the subscript $v = \{\mu, \sigma\}$ is used to indicate that the parameters are for mean and standard deviation, respectively. The RSS in LoS and NLoS links, as a function of path loss and antenna gain, can now be represented as:

$$R_l(h, r, B_{NS}, B_{EW}) = T + G(h, r, B_{NS}, B_{EW}) - 20 \log\left(\frac{4\pi f d}{c}\right) - X_l \tag{9.4}$$

$$R_n(h, r, B_{NS}, B_{EW}) = T + G(h, r, B_{NS}, B_{EW}) - 20 \log\left(\frac{4\pi f d}{c}\right) - X_n - X_s \tag{9.5}$$

where T is transmitted power, c is the speed of light, f denotes the frequency, d is the distance between UAV and MS and X_s is shadow fading Gaussian $\mathcal{N}\ (\mu_{sh}, \sigma_{sh})$ random variable (RV) with mean μ_{sh} and standard deviation σ_{sh}. For realistic system-level modelling of mobile systems, random components in dB, X_l and X_n are added as environment dependent variables utilizing the log-normal distribution with a zero mean [22]. X_l and X_n are therefore, $\mathcal{N}\ (0, \sigma_l)$ and $\mathcal{N}\ (0, \sigma_n)$ RVs, where σ_l and σ_n denote the standard deviations of X_l and X_n respectively. G is a function of $\phi_{MS}(h, r), B_{NS}$ and B_{EW} given by Equation (9.1).

9.5 UAV Coverage Model Development

Utilizing the system model in the preceding section, equations for coverage probability and the probability density functions of received signal strength as a function of key UAV parameters are derived in this section. This will enable us to analyze the resulting trade-offs between different design parameters in order to optimally design UAV-based cellular networks.

9.5.1 Coverage Probability

The RSS at the boundary exceeds a certain threshold, $\gamma = T - PL_{max}$ with a probability, $P_{cov} \geq \epsilon$, where PL_{max} is the maximum allowable path loss. We define this coverage probability, P_{cov} as:

$$P_{cov} = P[S_r \geq \gamma] = P_l(\phi_{MS})P[R_l(h, r, B_{NS}, B_{EW}) \geq \gamma]$$
$$+ P_n(\phi_{MS})P[R_n(h, r, B_{NS}, B_{EW}) \geq \gamma] \tag{9.6}$$

First, we calculate $P[R_n(h, r, B_{NS}, B_{EW}) \geq \gamma]$ using Equation (9.5) as in Equations (9.7)–(9.9), where Equation (9.9) is a result of complementary cumulative distribution function of a Gaussian random variable and substituting $\gamma = T - PL_{max}$ in Equation (9.8) and $G(\phi_{MS}, B_{NS}, B_{EW})$ is defined in Equation (9.1). Without loss of generality, we assume X_n and X_s to be independent RVs and thus $X'_n = X_n + X_s$ with mean $\mu'_n = \mu_{sh}$ and standard deviation, $\sigma'_n = \sqrt{\sigma^2_{sh} + \sigma^2_n}$.

$$P[R_n(h, r, B_{NS}, B_{EW}) \geq \gamma]$$

$$= P\left[T + G(\phi_{MS}, B_{NS}, B_{EW}) - 20\log\left(\frac{4\pi fd}{c}\right) - X_n - X_s \geq \gamma\right] \tag{9.7}$$

$$= \int_\gamma^\infty \frac{1}{\sqrt{2\pi}\sigma'_n} \exp\left[-\frac{1}{2}\left(\frac{X'_n - \left(T + G(\phi_{MS}, B_{NS}, B_{EW}) - 20\log\left(\frac{4\pi fd}{c}\right) - \mu'_n\right)}{\sigma'_n}\right)^2\right] dX'_n \tag{9.8}$$

$$= Q\left(\frac{20\log\left(\frac{4\pi fd}{c}\right) + \mu_{sh} - G(\phi_{MS}, B_{NS}, B_{EW}) - PL_{max}}{\sqrt{\sigma^2_{sh} + \sigma^2_n}}\right) \tag{9.9}$$

$Q(z)$ is the Q-function defined as: $Q(z) = \frac{1}{\sqrt{2\pi}} \int_z^\infty \exp(\frac{-u^2}{2}) du$.

Similarly, we compute $P[R_l(h, r, B_{NS}, B_{EW}) \geq \gamma]$, which yields the following expression:

$$P[R_l(h, r, B_{NS}, B_{EW}) \geq \gamma] = Q\left(\frac{20\log\left(\frac{4\pi fd}{c}\right) - G(\phi_{MS}, B_{NS}, B_{EW}) - PL_{max}}{\sigma_l}\right) \tag{9.10}$$

Moreover, from cell geometry in Figure 9.1 (under the assumption: height of MS $<< h$), ϕ_{MS} and d can be expressed as:

$$\phi_{MS} = \tan^{-1}\left(\frac{r}{h}\right), \quad d = \sqrt{h^2 + r^2} \tag{9.11}$$

For sake of simplicity of expression, we assume that the antenna beamwidth in EW and NS direction is symmetric i.e., $B_{EW} = B_{NS} = B$ in Equation (9.1) and ϕ_{tilt} in Figure 9.1 is zero unless otherwise stated. Note that with $\phi_{tilt} = 0$, the azimuth plane becomes perpendicular to the boresight the second part of Equation (9.1) is no longer applicable. Furthermore, G_{max} can be approximated as $10\log\left(\frac{29000}{B^2}\right)$ [20] and A_{max} can be ignored without impacting the required accuracy of this antenna model that mainly concerns gain on and around the boresight. Applying these simplifications to Equation (9.1), substituting Equation (9.1) and Equation (9.11) in Equations (9.9) and (9.10) and then making use of Equation (9.6) yields the expression for P_{cov} in Equation (9.12).

$$
P_{cov}(r, h, B) = \left(0.01j - \frac{0.01(j-k)}{1 + \left(\frac{\tan^{-1}\left(\frac{h}{r}\right)-l}{m} \right)^n} \right)
$$

$$
\times Q \left(\frac{20\log\left(\frac{4\pi f\sqrt{h^2+r^2}}{c}\right) - 10\log\left(\frac{29000}{B^2}\right) + 12\left(\frac{\tan^{-1}\left(\frac{r}{h}\right)-\phi_{tilt}}{B}\right)^2 - PL_{max}}{\sigma_l} \right)
$$

$$
+ \left(1 - 0.01j + \frac{0.01(j-k)}{1 + \left(\frac{\tan^{-1}\left(\frac{h}{r}\right)-l}{m} \right)^n} \right)
$$

$$
\times Q \left(\frac{20\log\left(\frac{4\pi f\sqrt{h^2+r^2}}{c}\right) + \mu_{sh} - 10\log\left(\frac{29000}{B^2}\right) + 12\left(\frac{\tan^{-1}\left(\frac{r}{h}\right)-\phi_{tilt}}{B}\right)^2 - PL_{max}}{\sqrt{\sigma_{sh}^2 + \sigma_n^2}} \right)
$$

$$(9.12)$$

Note that the radius r can also be translated into the position of MS on ground by: $r = \sqrt{x^2 + y^2}$, where x and y represent Cartesian coordinates for the location of user on ground.

9.5.2 Received Signal Strength

One way to investigate the RSS on a particular location on ground for a given height and antenna beamwidth is by evaluating the expected value of S_r over the random variables X_s, X_l and X_n as follows

$$
E[S_r] \overset{(k)}{=} P_l \, E[S_r|LoS] + P_n \, E[S_r|NLoS] = P_l \, E[R_l - R_n] + E[R_n]
$$

$$
= \left(0.01j - 1 - \frac{0.01(j-k)}{1 + \left(\frac{\tan^{-1}\left(\frac{h}{r}\right)-l}{m} \right)^n} \right) \mu_{sh} - 20\log\left(\frac{4\pi fd}{c}\right)
$$

$$
+ T - 12\left(\frac{\tan^{-1}\left(\frac{r}{h}\right) - \phi_{tilt}}{B} \right)^2 + 10\log\left(\frac{29000}{B^2}\right) \tag{9.13}
$$

where (k) is a result of the law of total expectation and P_l, R_l and R_n are functions of r, h and B and defined in Equations (9.2), (9.4) and (9.5) respectively.

However, since R_l and R_n are random variables due to shadowing, therefore, RSS can also be modelled as a random variable. In order to derive an analytical expression for the PDF of RSS at any arbitrary cell location, we express it as: $S_r = S_l + S_n$, where S_l and S_n are independent random variables, given by $S_l = P_l R_l$ and $S_n = P_n R_n$. The PDF of S_r is obtained by performing convolution of the two independent random variables. S_l and S_n and then applying some transformations of random variables. This results in the expression in Equation (9.14). Derivation of Equation (9.14) is provided in Appendix A.

$$f_{S_r}(s_r)$$

$$= \frac{\exp\left[\frac{-\left[s_r - \left(1 - 0.01j + \frac{0.01(j-k)}{1+\left(\frac{\tan^{-1}(h/r)-l}{m}\right)^n}\right)\left(\frac{p_\mu + \tan^{-1}(h/r)}{q_\mu + t_\mu \tan^{-1}(h/r)}\right) - T + 20\log\left(\frac{4\pi f(\sqrt{h^2+r^2})}{c}\right) - 10\log\left(\frac{29000}{B^2}\right) + 12\left(\frac{\tan^{-1}(r/h)}{B}\right)^2\right]^2}{2\sigma_l^2\left(\frac{0.01(j-k)}{1+\left(\frac{\tan^{-1}(h/r)-l}{m}\right)^n}\right)^2 + \left(1 - \frac{0.01(j-k)}{1+\left(\frac{\tan^{-1}(h/r)-l}{m}\right)^n}\right)^2\left(\sigma_n^2 + \left(\frac{p_\sigma + \tan^{-1}(h/r)}{q_\sigma + t_\sigma \tan^{-1}(h/r)}\right)^2\right)}\right]}{\sqrt{2\pi\sigma_l^2\left(\frac{0.01(j-k)}{1+\left(\frac{\tan^{-1}(\frac{h}{r})-l}{m}\right)^n}\right)^2 + \left(1 - \frac{0.01(j-k)}{1+\left(\frac{\tan^{-1}(\frac{h}{r})-l}{m}\right)^n}\right)^2\left(\sigma_n^2 + \left(\frac{p_\sigma + \tan^{-1}\left(\frac{h}{r}\right)}{q_\sigma + t_\sigma \tan^{-1}\left(\frac{h}{r}\right)}\right)^2\right)}}$$

$$(9.14)$$

The normalized PDF of received signal strength inside a geographical region (denoted by S) by assuming that the UAV resides at coordinates $(x, y, z) = (0, 0, h)$ where h is height of UAV, can then be found as follows:

$$f_S(s) = \frac{1}{A} \int\int_A f_{S_r}(s, x, y) dx dy \qquad (9.15)$$

in some geographical region A that lies in the xy-plane. The integral in Equation (9.15) can be solved through numerical methods.

9.6 Trade-Offs Between Coverage Radius, Beamwidth and Height

9.6.1 Coverage Radius Versus Beamwidth

As the height of UAV increases for a particular MS location, ϕ_{MS} in Equation (9.11) decreases (angle of elevation increases), leading to an increase in probability of LoS link in Equation (9.2), a decrease in shadowing in Equation (9.3) and an increase in free space path loss (as d increases). While the effect of these factors on coverage has been studied in earlier studies [7, 9, 10, 13, 15–18, 20, 22, 23], the impact of the fourth factor, antenna gain, in conjunction with these three factors remained unexamined. Figure 9.2(a) shows that the impact of this fourth factor is so profound that at a given cell radius, it results in a height (h'), after which the antenna gain trend with increasing beamwidth reverses. Figure 9.2(a) is plotted utilizing Equation (9.1) for $\phi_{tilt} = 0$ at $r = 5000$m under the assumptions stated in Section 9.6. The larger antenna gain at a given r is observed with increased height because, for the same MS location, ϕ_{MS} in Equation (9.11) decreases. For any two beamwidths, B_1 and B_2, h' is the point of

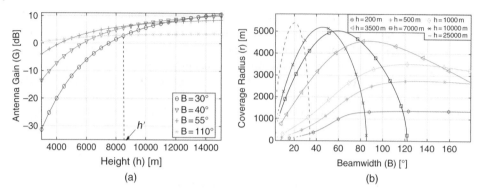

Figure 9.2 (a) Antenna gain at $r = 5000$ m with varying heights for different beamwidths. (b) Coverage radius against beamwidth for different heights.

intersection of the corresponding antenna gain versus height graphs in Figure 9.2a. As an example, $r = 5000$ m, $B_1 = 30°$ and $B_2 = 110°$ results in $h' = 8500$m as illustrated in Figure 1.2(a). h' can be calculated as follows:

$$10 \log \left(\frac{29000}{B_1^2} \right) - 12 \left[\frac{\tan^{-1} \left(\frac{r}{h'} \right)}{B_1} \right]^2 = 10 \log \left(\frac{29000}{B_2^2} \right) - 12 \left[\frac{\tan^{-1} \left(\frac{r}{h'} \right)}{B_2} \right]^2$$

$$h' = r \left(\tan \sqrt{\frac{5}{3} \log \left(\frac{B_2}{B_1} \right) \left(\frac{B_1^2 B_2^2}{B_2^2 - B_1^2} \right)} \right)^{-1} \tag{9.16}$$

The coverage radius versus beamwidth trend for different heights in Figure 9.2(b) can now be analyzed in light of the aforementioned factors. This figure is plotted by utilizing Equation (9.12) for PL_{max}=115dB, $\phi_{tilt} = 0°$, $\epsilon = 0.8$ and $f = 2$ GHz in a suburban environment. The trend shift in Figure 9.2(b) is attributed to the following: as height increases up to 1000m, the increase in antenna gain and decrease in shadowing offsets the increased free space path loss. As height increases beyond 1000m, the increase in antenna gain and decrease in shadowing is overshadowed by the increase in free space loss. As a result, the coverage radius increases with beamwidth, approaches to a maximum value and then starts to decrease. The analysis quantifies this maximum value of coverage radius in relation to antenna gain; for instance, maximum coverage radius of 5000m at a beamwidth of 55° for $h = 7000$m in Figure 9.2(b) can be attributed to the occurrence of largest antenna gain at a beamwidth of 55° at a height of 7000m in Figure 9.2(a).

9.6.2 Coverage Radius Versus Height

Figure 9.3(a) depicts the relation of coverage radius with height for different beamwidths. Initially, as the height of the UAV increases for small beamwidths, coverage radius also increases continuously. However, as beamwidth increases further, coverage radius versus height curves attain a parabolic shape. This is in

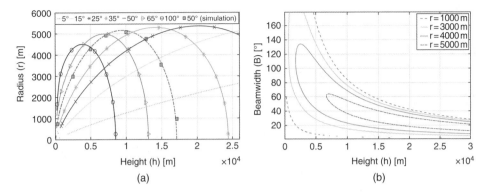

Figure 9.3 (a) Coverage radius against height for different beamwidths. (b) Beamwidth against height for different radii.

contrast to previous studies that suggest monotonic increase of UAV coverage radius with altitude [14, 19].

9.6.3 Height Versus Beamwidth

There can be scenarios where height of UAV is subject to changes due to factors beyond the system designer's control such as weather, but the same coverage pattern has to be maintained. The proposed model provides a mechanism to address such scenarios by characterizing the height versus beamwidth relationship in Figure 9.3(b). For example, if a UAV is deployed to cover $r_f = 4000$m with $\epsilon = 0.8$ at a height of 10,000m and if its height is changed to 9000m, the UAV will need to adjust its beamwidth from 70° to either 25° or 75° in order to continue providing the same coverage. Figure 9.3(b) also highlights the importance of optimizing both beamwidth and height in tandem with each other rather than independently.

9.7 Impact of Altitude, Beamwidth and Radius on RSS

In order to investigate the behaviour of RSS in context with UAV design parameters, Figure 9.4a and Figure 9.4b illustrate how the mean RSS varies with height and beamwidth with increasing cell radius r. In Figure 9.4a, mean RSS is plotted for $B = 50°$, $f = 2$GHz, $P_t = 40$dBm and $\phi_{tilt} = 0°$ in a suburban environment. As expected, RSS decreases as r increases for any fixed height. However, for any r, RSS initially increases with height, reaches to a peak value and then decreases as height increases further. The scenario in Figure 9.4b consists of a UAV deployed at a height of 3500m. Since a narrow beamwidth can only cover a small coverage area, the decrease in RSS with increasing radius is very rapid at low beamwidths. Hence, the trend of RSS with beamwidth is more clearly depicted in Figures 9.5a–9.5c, which are zoomed-in plots of Figure 9.4b for $20 \leq B < 180$. The continuous decrease of coverage radius with increase in beamwidth for $r < 800$m in Figure 9.5a can be attributed to the continuous decrease of antenna gain with radius in Figure 9.6 for $r < 800$m. As we approach closer to 1000m, the trend in antenna gain pattern starts to reverse for

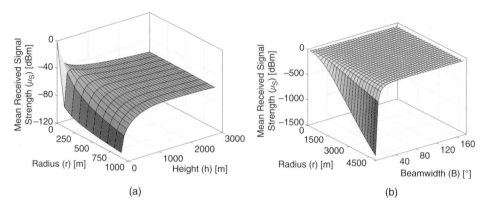

Figure 9.4 (a) Mean received signal with varying height for $B = 50°$. (b) Mean received signal with varying beamwidth at $h = 3500$m.

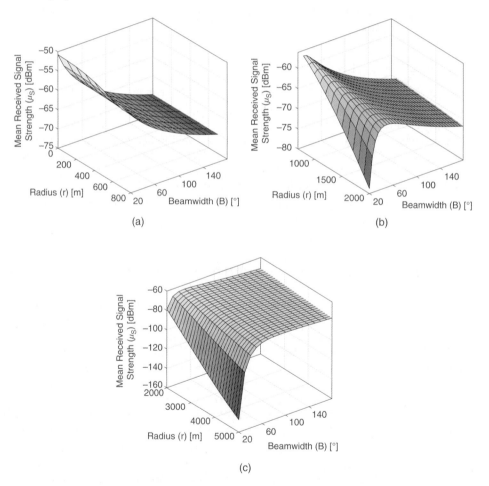

Figure 9.5 Mean received signal on ground with varying beamwidths $> 20°$. (a) 0m$< r <$ 800m; (b) 800m$< r <$ 2000m; (c) 2000m$< r <$ 5000m.

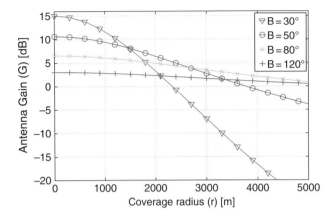

Figure 9.6 Antenna gain at h = 3500m with varying radius for different beamwidths.

different beamwidths, and we observe several intersecting points from $1000 < r < 2500$ in Figure 9.6 and hence mean RSS in Figure 9.5b attains a parabolic shape with increasing beamwidth in this range. At very large r, gain trend reverses completely and now it increases with increasing beamwidths, which is again in line with RSS trend in Figure 9.5c.

We now take a lot at the PDF of RSS inside a geographical area of 8000m×8000m. Normalized histograms of RSS from the simulation results and analytical PDFs from (9.15) are in agreement as shown in Figure 9.7–9.8. From Figure 9.7–9.8, we note that not only the range of RSS becomes narrower in a given area, but also the distribution of RSS approaches zero skewed Gaussian with either increasing height or increasing beamwidth.

9.8 Analysis for Different Frequencies and Environments

In Figure 9.9, we quantify the trade-off of coverage radius with beamwidth in changing environments and at different frequencies. From Figure 9.9, we observe that in addition to coverage radius decreasing sharply as frequency increases or environment becomes denser, the beamwidth at which this coverage radius versus beamwidth trend changes is also lower in a denser environment or at a higher frequency. For example, in a high-rise urban environment, decrease in radius starts from a beamwidth of as low as 30° at 2.0GHz, whereas for the same frequency, this decrease does not start until 110° in the case of the suburban environment. Similar observations can be made by observing the effect of frequency in the same environment. This is not just because of free space path loss, which increases with increasing frequency, but also because of shadowing, which increases at higher frequencies for the same environment [22]. In addition, the impact of frequency on coverage radius reduces as the environment becomes denser. These observations could play a valuable role for designing UAV-based cellular systems at higher frequencies by utilizing the unused part of higher frequency spectrum such as mmWave.

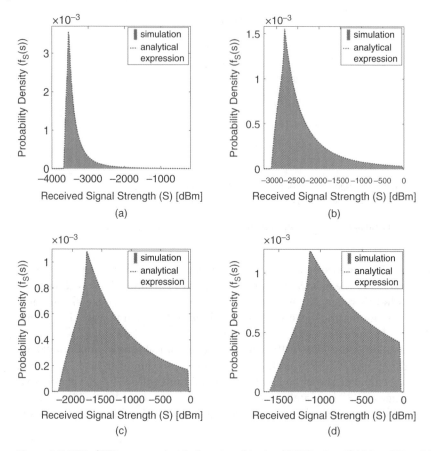

Figure 9.7 PDF of RSS on ground with changing altitude of UAV for $B = 5^o$ (a) $h = 500$m; (b) $h = 1500$m; (c) $h = 3500$m; (d) $h = 5500$m.

9.9 Comparison of Altitude and Beamwidth to Control Coverage

Noting that both beamwidth and height can be used for the same purpose of controlling coverage leads us towards comparing both of these parameters by analyzing the sensitivity of coverage radius to each of these parameters.

Figure 9.10 shows the gradient of radius with respect to beamwidth obtained by differentiating (9.12) with respect to beamwidth or determining gradient of the curves in Figure 9.2b with respect to beamwidth. By comparing the absolute values of $\Delta r / \Delta B$ in Figure 9.10, we conclude that the change in coverage radius is most sensitive to very high heights (up to 25,000m).

Next, we analyze the rate of change of radius with height ($\Delta r / \Delta h$) in Figure 9.10(c). The values of derivatives here follow a similar pattern with changing beamwidths except for beamwidths $< 25^o$ where $\Delta r / \Delta h$ is constant. By comparing values of the derivative in Figure 9.10(c) with Figure 9.10 (both comparisons range from samples of heights from between 0 to 25000m and beamwidths from 1° to 180°), we note that $\max(\Delta r / \Delta B) >>$

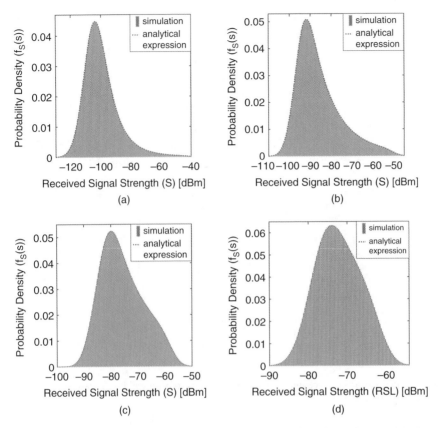

Figure 9.8 PDF of RSS on ground with changing altitude of UAV for $B = 50°$. (a) $h = 500$m; (b) $h = 1500$m; (c) $h = 3500$m; (d) $h = 5500$m.

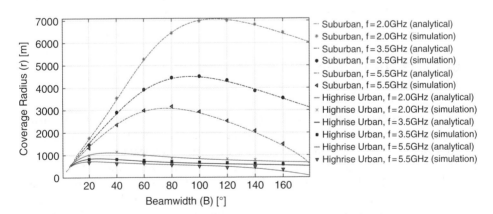

Figure 9.9 Coverage radius against beamwidth for varying frequency and environment at $h = 3000$m.

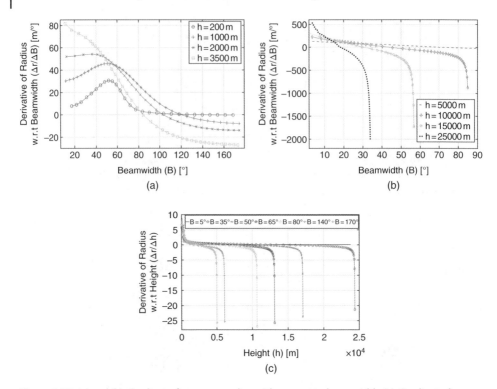

Figure 9.10 (a) and (b): Gradient of coverage radius with respect to beamwidth. (c): Gradient of coverage radius with respect to height.

$\max(\Delta r/\Delta h)$. This indicates that low beamwidths lead to greatest change (decrease) in coverage radius per unit beamwidth as compared to change in radius per unit height. This analysis can be leveraged to choose between the height and beamwidth or to design an appropriate combination of the two for optimizing coverage.

9.10 Coverage Probability with Varying Tilt Angles and Asymmetric Beamwidths

This section considers more flexible scenarios with varying tilt angles and asymmetrical beamwidths in north-south and east-west directions. In Figure 9.11(a)–(b), we compare the probability of coverage using zero antenna tilt with a tilt angle of 10°. For non-zero tilt angles, the coverage pattern forms an off-centred ellipse shape rather than a circle centred at the origin. The coverage probability in case of non-zero tilt angle is evaluated by directly substituting (9.2), (9.9) and (9.10) in (9.6) without any assumptions regarding the coverage pattern. Figure 9.11(c)–(d) shows how the coverage probability changes with different values of B_{NS} and B_{EW}. The effect of changing tilt values is depicted in Figure 9.12. Although a UAV with antenna tilt of 80° covers more area as compared to tilt angle of 20°, the maximum coverage probability with $\phi_{tilt} = 80°$ is reduced by half as compared to $\phi_{tilt} = 20°$.

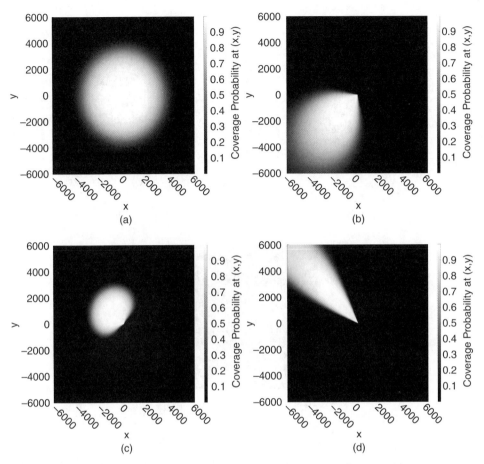

Figure 9.11 (a) and (b): Coverage probability at $B = 50°$ and $h = 5000$m with symmetrical beamwidths; (c) and (d): Coverage probability with asymmetrical beamwidths at $\phi_{tilt} = 10°$, $\theta_a = -45°$ and $h = 5000$m. (a) $\phi_{tilt} = 0°$; (b) $\phi_{tilt} = 10°$, $\theta_a = 45°$; (c) $B_{NS} = 20°$, $B_{EW} = 70°$; (d) $B_{NS} = 70°$, $B_{EW} = 20°$

9.11 Coverage Analysis with Multiple UAVs

Previous literature [19], utilizes circle packing theory to determine the number of UAVs to achieve a gain in coverage probability in a certain geographical area. In the circle packing problem, N identical circles (cells) are arranged inside a larger circle (target area) of radius R_t such that the packing density is maximized and none of the circles overlap [23]. The radius of each of the N circles that solves this problem is denoted by r_{max} and one UAV provides coverage to one small cell (circle). However, this approach towards determining the needed number of UAVs for achieving a coverage level has two major drawbacks. Firstly, significant gaps between circles or cells are inevitable when two or more circles are used to cover a given area. This is due to the inherent nature of circle packing theory, since in order to cover the target area completely, $N \to \infty$ and $r_{max} \to 0$. Secondly, the number of circles (UAVs) increase rapidly with desired coverage

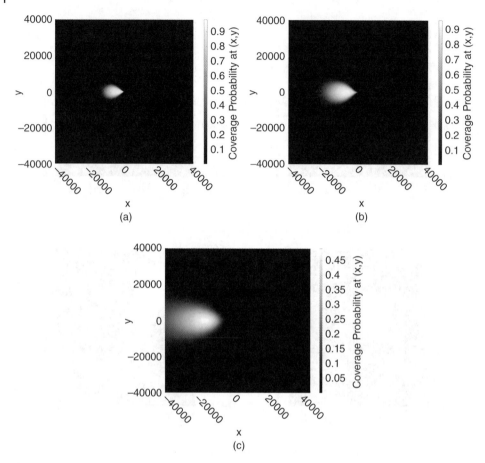

Figure 9.12 Coverage probability with varying tilt angles at $B = 40°$ and $h = 10,000$m. (a) $\phi_{tilt} = 20°$; (b) $\phi_{tilt} = 40°$; (c) $\phi_{tilt} = 80°$.

probability. We overcome both of these problems by introducing a UAV placement model that simply uses hexagonal cell shapes instead of circle. This approach not only resolves aforementioned problems but also leads to a better coverage. To illustrate our approach, consider the case for $N = 3$ in Figure 9.13. If we consider a hexagonal area with the longest distance from centre to the edge denoted by R_t and the distance from centre to the vertex of a hexagon by r_h, then the maximum distance between any two farthest hexagons will be $4r_h$ as shown in Figure 9.13b. Our goal is to minimize this distance in order to maximize the packing density, thus leading to the arrangement shown in Figure 9.13c. Here, the maximum distance between farthest hexagons is $4r_h \sqrt{3}/2 < 4r_h$, leading to $r_h = 0.5R_t = r_{max}$. The maximum total coverage (in percentage) for $N = 3$ can then be calculated as follows:

$$C_h = \frac{\text{area covered by 3 UAVs}}{\text{total area to be covered}} = \frac{3\left(\frac{3}{2}\sqrt{3}\left(\frac{1}{2}R_t\right)^2\right)}{\frac{3}{2}\sqrt{3}R_t^2} = 75\% \qquad (9.17)$$

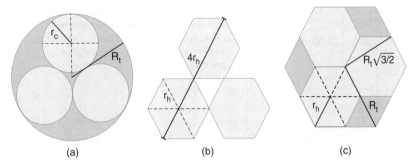

Figure 9.13 Circle packing versus hexagonal packing for $N = 3$. (a) Circle packing (b) Max. distance between hexagons (c) Hexagon packing.

Table 9.2 Comparison between circular and hexagonal packing in terms of coverage radius of each UAV and maximum total coverage.

N	r^c_{max} [19]	r^h_{max}	C_c [19]	C_h
1	R_t	R_t	1	1
2	$0.500\ R_t$	$0.447\ R_t$	50.0	75.0
3	$0.464\ R_t$	$0.500\ R_t$	64.6	75.0
4	$0.413\ R_t$	$0.400\ R_t$	68.6	75.0
5	$0.370\ R_t$	$0.333\ R_t$	68.5	75.0
6	$0.333\ R_t$	$0.286\ R_t$	66.6	75.0
7	$0.333\ R_t$	$0.333\ R_t$	77.8	77.8
8	$0.302\ R_t$	$0.286\ R_t$	73.3	72.7
9	$0.275\ R_t$	$0.250\ R_t$	68.9	69.2
10	$0.261\ R_t$	$0.286\ R_t$	68.7	75.0

Similar analysis is done for $N = 1$ to 10 and presented in Table 9.2, where C_c and C_h represent the maximum percentage of actual area covered out of total target area using circular and hexagonal packing respectively while r^c_{max} and r^h_{max} represent the maximum possible radius of each cell using circular and hexagonal packing approaches respectively. We compare the results of our proposed approach with those from [19] that utilize circle packing to the same effect. From our proposed approach, the minimum possible coverage is ~70 % for any number of UAVs, whereas with circle packing theory it drops to as low as 50% [19]. This has a direct impact on the coverage threshold requirement of the system. It is highlighted in [19] that a 0.7 coverage performance is impossible to achieve with $1 < N < 7$ using a circle packing approach. Our proposed hexagonal packing strategy, on the other hand, ensures that this coverage performance demand can be met with a much smaller number of UAVs. We illustrate this by calculating the minimum number of UAVs required to cover different geographical areas by utilizing Table 9.2 for a coverage threshold $\geq 70\%$, with a tolerance of $\pm 1\%$ for $\epsilon = 0.8$ over $0 < h \leq 5000$ and $1 < B < 180$. Figure 9.14(a) compares the resulting minimum number of UAVs obtained with hexagonal packing and circle packing from [19]. For $C = 70\%$, from circle packing approach, we can cover a desired area up to 14 km with 1, 7 or 8 UAVs. On the

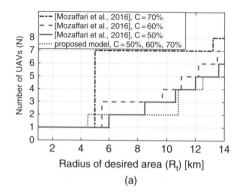

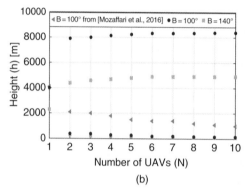

Figure 9.14 (a) Minimum number of UAVs versus radius of desired area for different minimum coverage thresholds. (b) Altitude with varying no. of UAVs for different beamwidths.

other hand, with our proposed hexagonal packing approach, we can cover an area with a much smaller number of UAVs, i.e., 1, 2, 4 or 5 UAVs. Another advantage that hexagonal packing offers is the relative scalability of number of UAVs as the coverage threshold changes, which in case of circle packing increases rapidly as coverage threshold goes from 50% to 80% as seen in Figure 9.14(a).

Next, we investigate the relationship between number of UAVs and beamwidth of each UAV. First, we find the maximum possible r for a given geographical area with radius R_t using Table 9.2 as the number of UAVs vary and then the corresponding beamwidth using Figure 9.2b from our proposed model.

Finally, in order to observe the trend of UAV altitude as the number of UAVs vary, the optimal UAV altitude that yields maximum possible coverage is plotted in Figure 9.14(b) for different number of UAVs. The system model using binary antenna gain pattern proposed in earlier studies such as [19] does not reflect the role of beamwidth in UAV altitude with increasing number of UAVs. Hence, UAV altitude decreases monotonically as number of UAVs increases. Figure 9.14(b) shows this is not the case when a practical antenna gain pattern, as proposed in this study, is used. In the plot, one UAV covers maximum possible area for a certain beamwidth that can be found from Figure 9.3(a). For beamwidths up to 15°, UAV altitude with number of UAVs follow the same trend as in [19]. However, the trend changes for higher beamwidths. This trend is explicitly compared with the model in [19] for a beamwidth of 100°. This concludes that, contrary to observation made in prior studies with simple or no antenna models, it is not necessary for UAV altitude to decrease monotonically as the number of UAV increases; in fact, it can also either increase monotonically or behave as a combination of increasing and decreasing altitude as the number of UAVs increase.

9.12 Conclusion

This chapter provides a holistic analysis of the interplay between key UAV deployment parameters: coverage radius, height and beamwidth while considering design space dimensions that remain unexplored in existing studies. It further provides a mathematical model to estimate RSS at any distance from boresight of antenna as a function of antenna beamwidth and altitude. The analysis and results provide several new insights that prior models with no or simplified antenna, path loss, or shadowing models do not reveal, such as:

- UAV altitude or antenna beamwidth does not have to necessarily increase continuously for higher coverage radius. There exists an optimal UAV altitude and antenna beamwidth for maximum coverage.
- Contrary to findings reported in some prior studies, UAV coverage radius does not necessarily increase as altitude increases.
- The minimum number of UAVs required to cover a given area does not necessarily decrease monotonically as UAV altitude increases. These results allow us to determine optimal UAV parameters for realistic deployment.
- Antenna beamwidth and altitude should be optimized simultaneously rather than independently as is the case assumed in previous works.
- Optimizing beamwidth instead of height to control coverage may be a more practical and that coverage is most sensitive to beamwidths of less than 40°.
- A hexagonal packing scheme for solving coverage optimization problem with multiple UAVs leads to better coverage as it leaves much smaller coverage holes. Thus it can cover a higher proportion of the given area with same number of UAVs compared to circle packing and is scalable in terms of number of UAVs with increasing probability of coverage.

Acknowledgment

This material is based upon work supported by the National Science Foundation under Grant Number 1619346 and Grant Number 1718956.

References

1 Ali Imran and Rahim Tafazolli. Performance & capacity of mobile broadband WiMAX (802.16 e) deployed via high altitude platform. In *IEEE European Wireless Conference*, pages 319–323, 2009.

2 Y. Zeng, R. Zhang, and T. J. Lim. Wireless communications with unmanned aerial vehicles: opportunities and challenges. *IEEE Communications Magazine*, 54(5):36–42, 2016.

3 M. Kamel, W. Hamouda, and A. Youssef. Ultra-dense networks: A survey. *IEEE Communications Surveys Tutorials*, 18(4):2522–2545, 2016.

4 W. H. Robinson and A. P. Lauf. Resilient and efficient MANET aerial communications for search and rescue applications. In *International Conference on Computing, Networking and Communications (ICNC)*, pages 845–849, 2013.

5 S. Kandeepan, K. Gomez, L. Reynaud, and T. Rasheed. Aerial-terrestrial communications: terrestrial cooperation and energy-efficient transmissions to aerial base stations. *IEEE Transactions on Aerospace and Electronic Systems*, 50(4):2715–2735, October 2014.

6 Jiangbin Lyu, Yong Zeng, and Rui Zhang. UAV-aided cellular offloading: a potential solution to hot-spot issue in 5G. *submitted to IEEE Trans. Wireless Commun*, 2017.

7 D. Grace, J. Thornton, Guanhua Chen, G. P. White, and T. C. Tozer. Improving the system capacity of broadband services using multiple high-altitude platforms. *IEEE Transactions on Wireless Communications*, 4(2):700–709, March 2005.

8 Ali Imran, Majid Shateri, and Rahim Tafazolli. On the comparison of performance, capacity and economics of terrestrial base station and high altitude platform based deployment of 4G. In *Proceedings of the 6th ACM Symposium on Performance Evaluation of Wireless Ad Hoc, Sensor, and Ubiquitous Networks*, pages 58–62, 2009.

9 A. Al-Hourani, S. Kandeepan, and S. Lardner. Optimal LAP altitude for maximum coverage. *IEEE Wireless Communications Letters*, 3(6): 569–572, 2014.

10 A. Al-Hourani, S. Chandrasekharan, G. Kaandorp, W. Glenn, A. Jamalipour, and S. Kandeepan. Coverage and rate analysis of aerial base stations [letter]. *IEEE Transactions on Aerospace and Electronic Systems*, 52(6):3077–3081, December 2016.

11 M. Mozaffari, W. Saad, M. Bennis, and M. Debbah. Drone small cells in the clouds: Design, deployment and performance analysis. In *IEEE Global Communications Conference (GLOBECOM)*, pages 1–6, 2015.

12 J. Thornton, D. Grace, M. H. Capstick, and T. C. Tozer. Optimizing an array of antennas for cellular coverage from a high altitude platform. *IEEE Transactions on Wireless Communications*, 2(3):484–492, May 2003.

13 N. Goddemeier, K. Daniel, and C. Wietfeld. Coverage evaluation of wireless networks for unmanned aerial systems. In *2010 IEEE Globecom Workshops*, pages 1760–1765, Dec 2010.

14 J. Košmerl and A. Vilhar. Base stations placement optimization in wireless networks for emergency communications. In *IEEE International Conference on Communications Workshops (ICC)*, pages 200–205, 2014.

15 R. I. Bor-Yaliniz, A. El-Keyi, and H. Yanikomeroglu. Efficient 3-D placement of an aerial base station in next generation cellular networks. In *IEEE International Conference on Communications*, pages 1–5, 2016.

16 Akram Al-Hourani and Sithamparanathan Kandeepan. Cognitive relay nodes for airborne LTE emergency networks. In *7th International Conference on Signal Processing and Communication Systems (ICSPCS)*, pages 1–9, 2013.

17 J. Holis, D. Grace, and P. Pechac. Effect of antenna power roll-off on the performance of 3G cellular systems from high altitude platforms. *IEEE Transactions on Aerospace and Electronic Systems*, 46(3):1468–1477, July 2010.

18 Haiyun He, Shuowen Zhang, Yong Zeng, and Rui Zhang. Joint altitude and beamwidth optimization for UAV-enabled multiuser communications. *IEEE Communications Letters*, 22(2):344–347, 2018.

19 M. Mozaffari, W. Saad, M. Bennis, and M. Debbah. Efficient deployment of multiple unmanned aerial vehicles for optimal wireless coverage. *IEEE Communications Letters*, 20(8):1647–1650, 2016.

20 Constantine A Balanis. *Antenna Theory: Analysis and Design*. John Wiley & Sons, 2016. ISBN 1118642066.

21 3rd Generation Partnership Project. TR36.814 v9.0.0: Further advancements for E-UTRA physical layers aspects (release 9). Technical report, 3GPP, Sophia Antipolis, France, March 2010.

22 J. Holis and P. Pechac. Elevation dependent shadowing model for mobile communications via high altitude platforms in built-up areas. *IEEE Transactions on Antennas and Propagation*, 56(4):1078–1084, 2008.

23 Zsolt Gáspár and Tibor Tarnai. Upper bound of density for packing of equal circles in special domains in the plane. *Periodica Polytechnica. Civil Engineering*, 44(1):13, 2000.

Appendix A
Derivation of Equation (9.14)

The probability density function of RSS at an arbitrary point in a cell is found by first deriving PDFs of LoS and NLoS component of RSS using transformations of RVs. Thereafter, Theorem 1 is applied.

From Section 9.5.2, we can express the RV, S_l as:

$$S_l = P_l\, R_l$$

$$= P_l \left[10\log\left(\frac{29000}{B^2}\right) - 12\left(\frac{\tan^{-1}\left(\frac{r}{h}\right) - \phi_{tilt}}{B}\right)^2 + T - 20\log\left(\frac{4\pi fd}{c}\right) \right] - P_l X_l$$

$$= P_l A_1 - P_l X_l \tag{A.1}$$

where P_l is given in (9.2) and A_1 can be treated as a constant for a UAV deployed at a fixed height and beamwidth, given by:

$$A_1 = T + 10\log\left(\frac{29000}{B^2}\right) - 12\left(\frac{\tan^{-1}\left(\frac{r}{h}\right) - \phi_{tilt}}{B}\right)^2 - 20\log\left(\frac{4\pi fd}{c}\right) \tag{A.2}$$

We proceed by first finding PDF of S_l by applying transformations of random variables as follows:

$$F_{S_l}(s_l) = P(S_l \leq s_l)$$

$$= P(P_l A_1 - P_l X_l \leq s_l)$$

$$= P\left(X_l \leq \frac{s_l - P_l A_1}{P_l}\right)$$

$$F_{S_l}(s_l) = F_{X_l}\left(\frac{s_l - P_l A_1}{P_l}\right) \tag{A.3}$$

where $F_{S_l}(s_l)$ and $F_{X_l}(x_l)$ are the cumulative distribution functions (CDFs) of S_l and X_l respectively. Both sides of (A.3) are a function of s_l and therefore we differentiate both sides w.r.t s_l in order to get the PDF:

$$f_{S_l}(s_l) = f_{X_l}\left(\frac{s_l - P_l A_1}{P_l}\right) \frac{d}{ds_l}\left(\frac{s_l - P_l A_1}{P_l}\right)$$

$$f_{S_l}(s_l) = \frac{1}{P_l}f_{X_l}\left(\frac{s_l - P_l A_1}{P_l}\right) \tag{A.4}$$

This allows us to find the PDF of S_l based on the PDF of X_l which is $\mathcal{N}(0, \sigma_n)$ random variable. Applying the transformation in (A.4) yields the following expression for $f_{S_l}(s_l)$:

$$f_{S_l}(s_l) = \frac{\exp\left(-\frac{(s_l - P_l A_1)^2}{2(P_l \sigma_l)^2}\right)}{\sqrt{2\pi}P_l \sigma_l} \tag{A.5}$$

Following a similar procedure, we then derive PDF of S_n, which yields following expression:

$$f_{S_n}(s_n) = \frac{\exp\left(-\frac{(s_n - [P_n(A_1 - \mu_{sh})])^2}{2(P_n)^2(\sigma_n^2 + \sigma_{sh}^2)}\right)}{\sqrt{2\pi}P_n\sqrt{\sigma_n^2 + \sigma_{sh}^2}} \tag{A.6}$$

We can now proceed to apply Theorem 1 to derive PDF of S_r by performing convolution of (A.5) with (A.6) as in (A.8), where A_2 in (A.8) is expanded in (A.9).

$$f_{S_r}(s_r) = \int_{-\infty}^{\infty} \frac{1}{\sqrt{2\pi}\,P_n\sqrt{\sigma_n^2 + \sigma_{sh}^2}} \exp\left[-\frac{(s_r - s_l - [P_n(A_1 - \mu_{sh})])^2}{2(P_n)^2(\sigma_n^2 + \sigma_{sh}^2)}\right]$$

$$\times \frac{1}{\sqrt{2\pi}\,P_l\,\sigma_l} \exp\left[-\frac{(s_l - P_l A_1)^2}{2(P_l\,\sigma_l)^2}\right] ds_l \tag{A.7}$$

$$= \int_{-\infty}^{\infty} \frac{1}{\sqrt{2\pi}\sqrt{2\pi}\,P_n\sqrt{\sigma_n^2 + \sigma_{sh}^2}\,P_l\,\sigma_l} \exp\left[-\frac{A_2}{2(P_l)^2\sigma_l^2(P_n)^2(\sigma_n^2 + \sigma_{sh}^2)}\right] ds_l \tag{A.8}$$

$$\begin{aligned}
A_2 &= (P_l)^2\,\sigma_l^2\,(s_r - s_l - [P_n(A_1 - \mu_{sh})])^2 + (P_n)^2(\sigma_n^2 + \sigma_{sh}^2)(s_l - P_l A_1)^2 \\
&= (P_l)^2\,\sigma_l^2\,(s_r^2 + s_l^2 + [P_n(A_1 - \mu_{sh})]^2 - 2s_l s_r - 2s_r[P_n(A_1 - \mu_{sh})] \\
&\quad + 2s_l[P_n(A_1 - \mu_{sh})]) + (P_n)^2(\sigma_n^2 + \sigma_{sh}^2)(s_l^2 + (P_l A_1)^2 - 2s_l P_l A_1) \tag{A.9} \\
A_2 &= s_l^2[(P_l)^2\sigma_l^2 + (P_n)^2(\sigma_n^2 + \sigma_{sh}^2)] - 2s_l[(P_l \sigma_l)^2(s_r - P_n(A_1 - \mu_{sh})) \\
&\quad + (P_n)^2(\sigma_n^2 + \sigma_{sh}^2)P_l A_1] + (P_l \sigma_l)^2(s_r^2 + [P_n(A_1 - \mu_{sh})]^2 - 2s_r[P_n(A_1 - \mu_{sh})]) \\
&\quad + (P_n)^2(\sigma_n^2 + \sigma_{sh}^2)[P_l A_1]^2 \tag{A.10}
\end{aligned}$$

Next, we perform algebraic manipulation on (A.9) in (A.10) to convert the terms in it to $(C + D)^2$ form in order to apply completing the squares method. In order to complete the square of (A.10), we define a new variable in the following manner:

$$\sigma_s = \sqrt{\sigma_{s_l}^2 + \sigma_{s_n}^2} \tag{A.11}$$

where $\sigma_{s_l}^2 = (P_l\,\sigma_l)^2$ and $\sigma_{s_n}^2 = (P_n)^2(\sigma_n^2 + \sigma_{sh}^2)$.

Therefore, our PDF expression in (27) reduces to:

$$f_{S_r}(s_r) = \int_{-\infty}^{\infty} \frac{1}{\sqrt{2\pi}\sigma_s} \frac{1}{\sqrt{2\pi}\frac{\sigma_{s_l}\sigma_{s_n}}{\sigma_s}} \exp\left[-\frac{A_3}{2(\frac{\sigma_{s_l}\sigma_{s_n}}{\sigma_s})^2}\right] \tag{A.12}$$

where A_3 equals (A.13).

$$A_3 = s_l^2 - 2s_l\frac{\sigma_{s_l}^2(s_r - [P_n(A_1 - \mu_{sh})]) + \sigma_{s_n}^2 P_l A_1}{\sigma_s^2}$$
$$+ \frac{\sigma_{s_l}^2(s_r^2 + [P_n(A_1 - \mu_{sh})]^2 - 2s_r[P_n(A_1 - \mu_{sh})] + \sigma_{s_n}[P_l A_1]^2}{\sigma_s^2} \tag{A.13}$$

Squares are now completed by collecting the appropriate terms as in (A.14). Finally, the exponent in (A.12) can be broken into a product of two exponents as shown in (A.15). By noting that the integral in (A.15) is in fact a Gaussian distribution on S_l (and hence integrates to 1), substituting σ_s from (A.11), P_l and P_n from (A.2) leads to the expression of PDF of S_r in (A.14).

$$A_3 = \left(s_l - \frac{\sigma_{s_l}^2(s_r - [P_n(A_1 - \mu_{sh})]) + \sigma_{s_n}^2 P_l A_1}{\sigma_s^2}\right)^2$$
$$- \left(\frac{\sigma_{s_l}^2(s_r - [P_n(A_1 - \mu_{sh})]) + \sigma_{s_n}^2 P_l A_1}{\sigma_s^2}\right)^2$$
$$+ \left(\frac{\sigma_{s_l}^2(s_r - [P_n(A_1 - \mu_{sh})])^2 + \sigma_{s_n}^2 P_l^2 A_1^2}{\sigma_s^2}\right) \tag{A.14}$$

$$f_{S_r}(s_r) = \frac{1}{\sqrt{2\pi}\sigma_s} \exp\left[-\frac{(s_r - [P_l A_1 + P_n(A_1 - \mu_{sh})])^2}{2\sigma_s^2}\right]$$

$$\times \underbrace{\int_{-\infty}^{\infty} \frac{1}{\sqrt{2\pi}\frac{\sigma_{s_l}\sigma_{s_n}}{\sigma_s}} \exp\left[-\frac{s_l - \frac{\sigma_{s_l}^2(s_r - [P_n(A_1 - \mu_{sh})]) + \sigma_{s_n}^2[P_l A_1]}{\sigma_s^2}}{2(\frac{\sigma_{s_l}\sigma_{s_n}}{\sigma_s})^2}\right] ds_l}_{=1} \tag{A.15}$$

10

Intelligent Positioning of UAVs for Future Cellular Networks

João Pedro Battistella Nadas, Paulo Valente Klaine, Rafaela de Paula Parisotto and Richard D. Souza

10.1   Introduction

One of the most promising paradigms for the fifth generation of cellular systems (5G) is the concept of self-organizing networks (SON) [1]. SON consist of the network monitoring its own performance, through the collection and measurement of real-time data, detecting changes in patterns or abnormalities and reacting to these changes accordingly. For this to be possible, however, a high level of automation is required, so that future networks can adapt themselves without any interaction from human operators. Furthermore, SON are also expected to be extremely agile, in order to maintain their desired objectives and deliver the best possible service to end users. In addition, the spur of novel applications and the advanced capabilities of modern handheld devices also creates a demand for flexible and intelligent solutions, and thus it can be said that SON will play a vital role in future cellular networks [1].

In order to meet the stringent requirements of SON and fully enable its concept, the use of unmanned aerial vehicles (UAV) equipped with radio devices has a rich potential [2]. UAVs can provide the required adaptability, by moving themselves around, and agility, as they can be easily deployed and integrated into the existing infrastructure. However, determining the best position to deploy such UAVs is a challenging task and it is still an open question in the research community [2]. One possible way is to place the UAVs manually, or according to certain patterns, depending on user distribution and requirements [3]. However, this type of solution would not fully conform with the characteristics of SON, as they would require extensive manual planning. Thus, in order to fully enable the concept of SON while integrating UAVs into future cellular networks, it is clear that these flying platforms need to be equipped with some intelligence, which can be achieved by the use of machine learning (ML) algorithms [4].

The objective of this chapter is to motivate the use of intelligent algorithms to determine the best position of UAVs which can be used to enhance the capabilities of existing cellular networks. For instance, UAVs can improve the capacity of current urban deployments, or provide connectivity where the existing network cannot, such as in remote/rural areas and in the case of natural disasters [2]. Therefore, we present an intelligent positioning algorithm based on reinforcement learning (RL), originally proposed in [4], and evaluate its performance in an emergency communication network

Enabling 5G Communication Systems to Support Vertical Industries, First Edition.
Edited by Muhammad Ali Imran, Yusuf Abdulrahman Sambo and Qammer H. Abbasi.

(ECN) scenario. The remainder of this chapter is organized as follows: Section 10.2 presents future applications of UAVs in cellular networks; Section 10.3 presents the state-of-the-art of positioning systems for UAVs in communication networks; Section 10.4 contains a brief summary of how RL works. Section 10.5 presents an intelligent positioning algorithm and results of simulations in an ECN scenario; Section 10.6 concludes the chapter.

10.2 Applications of UAVs in Cellular Networks

Due to their flexibility, UAVs offer many opportunities in cellular networks to provide service where today it is not currently practical, or to enhance the quality of experience (QoE) of end users. By using UAVs it is possible to extend both radio access network (RAN) and backhaul capabilities of current networks in novel and inventive ways [2].

For instance, aerial platforms can be used to enhance mobile broadband via caching [5] or relaying, serving as a flying fronthaul [6], providing additional network radio access capacity during events or unexpected crowds (pop-up networks), deploying emergency communication networks for disaster scenarios [4], and many other applications. Next, we explore in more detail a few possible applications that UAVs can enable in future cellular networks.

10.2.1 Coverage in Rural Areas

Despite the ubiquitous mobile coverage in most urban areas, there are still various rural communities which face challenges in getting connected. The primary reason for this is the low cost-effectiveness of installing fixed cellular networks in such locations, as they are often far from big centres and have sparse population [7]. The elevated cost of installing a fixed infrastructure, combined with the previously mentioned points, makes rural areas unattractive to mobile operators, creating a vacuum of coverage.

Given that UAVs are much more inexpensive and flexible to deploy, using them to provide connectivity in rural areas is one potentially disruptive application, which can connect isolated communities at a cost-effective price. For instance, intelligently positioned high-altitude platforms could serve as base stations (BS) in rural areas. These flying platforms could be deployed in either a multi-tier manner, in which the lower tiers are responsible for providing RAN connectivity to end users and the higher tiers are responsible for backhaul and inter-UAV connections [6] or in a single tier and perform the backhaul connection via wireless or satellite links.

10.2.2 Communication for Internet of Things

Current cellular networks have been designed with a focus on serving handheld devices. However, novel applications based on Internet of things (IoT) have much more diverse requirements and thus can be served efficiently by networks designed with those requirements in mind [8].

For example, in a wireless sensor network (WSN) nodes must be energy-efficient as they are typically battery-operated [9]. WSNs could be used, for instance, in smart city infrastructure management, healthcare monitoring, smart transport, smart grids and many other futuristic applications [2]. Furthermore, the batteries on the nodes may be non-replaceable or rechargeable, meaning that when it uses its charge, a new node must be deployed in its place. In this scenario, having an energy-efficient communication with low transmit power in the uplink is essential. However, in traditional mobile network deployments, BSs can be located hundreds of meters or even kilometres from the end user device, which requires high transmit power on the uplink of the sensor nodes, or the transmission of the message through multiple hops [10].

Given this, UAVs could be used as mobile BSs which can fly close to the IoT nodes and collect the information with low uplink energy consumption. Moreover, designing the trajectory of UAVs to be efficient can be achieved using ML algorithms. Another possible application involves the intelligent positioning of UAVs in areas covered by IoT sensors, such as rural areas. In this case, a trade-off between coverage and uplink transmission power is an interesting problem to be investigated, as being near the sensors is advantageous to reduce the uplink transmission power (and maximize the sensor's lifetime), but at the expense of covering fewer sensors.

10.2.3 Flying Fronthaul/Backhaul

Ultra-dense networks comprised of macro BSs and various small cell BSs are considered one of the key enablers of 5G [6]. However, connecting the various small cell BSs to the core network is an expensive endeavour and providing optical fibre fronthaul/backhaul connection may not be viable in all deployments. Thus, one feasible alternative is the use of UAVs with free-space optics or highly directional mmWave communication to connect the small cell BSs with the core network.

Early results by [6] show that using UAVs in an adequate architecture can provide excellent performance and is an attractive cost-effective solution for the fronthaul/backhaul connection of small cell BSs. Another possible application involves the use of flying platforms as a backhaul solution for delay-tolerant applications, such as after a disaster has happened or the retrieval of information from sensors in remote areas. In this case UAVs could fly around the area collecting the information and then return, delivering all the collected information to the core network.

10.2.4 Aerial Edge Caching

Edge content caching consists of storing the most requested contents close to end users to offload the traffic from the core network and improve user experience alongside overall network performance. However, due to user mobility, edge caching solutions often require the contents to be stored in multiple locations, limiting its applicability [2].

Cache-enabled UAV BSs combined with user mobility prediction can dynamically cache content and move to where the contents are most needed. In this way, users can be served efficiently. Moreover, the UAVs can fly to locations where the content is pre-cached and reduce even further the burden on the core network.

10.2.5 Pop-Up Networks

The exponential growth in the number of mobile devices and digital services in the recent years has created a massive demand for people to be connected at all times. Furthermore, the capabilities of modern devices coupled with the complexity of services offered creates challenges for mobile operators to provide the QoE required by users [11].

Typically, cellular deployments have to be pre-configured to cope with peaks in traffic, which means that operators often have to design the whole network with the worst-case scenario in mind, which is extremely inefficient [1]. However, even if the operators manage to achieve a good design, sometimes – e.g. when big events happen or when there is a sudden agglomeration of people in any region of a city – the network cannot cope with the increase in demand. In the context of SON, one interesting strategy is the deployment of pop-up networks, which would operate when the current network cannot handle the requirements of users, providing a variable level of coverage which can scale with the demand. In addition, since it is hard to predict where these networks would be needed, fixed solutions are not ideal. Instead, a viable alternative is to use intelligent UAVs to provide additional RAN resources efficiently.

10.2.6 Emergency Communication Networks

Consider a scenario where a disaster has occurred and disrupted the existing cellular network. In this case establishing a quick temporary network is crucial to locate and rescue people who are in the affected area. Because of their flexibility and ease of deployment, UAVs are a viable solution to create an emergency communication network (ECN) [4].

We envision using multiple UAVs to create a provisional network as depicted in Figure 10.1, where UAVs provide coverage to certain users and a truck-mounted BS can be used to provide connectivity between the UAVs and the core network (backhaul) as well as to nearby users. This scenario is further explored in Section 10.5, where we present one proposed approach for positioning UAVs in ECNs.

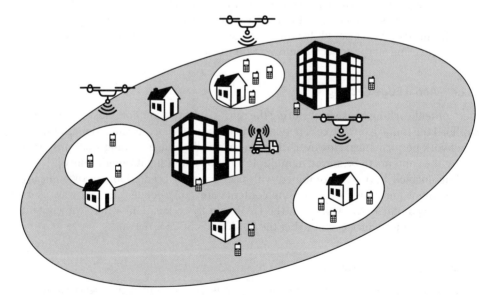

Figure 10.1 The envisioned scenario, with a truck-mounted base station to provide connectivity to the backhaul and UAVs to create a provisional network.

10.3   Strategies for Positioning UAVs in Cellular Network

All the applications in Section 10.2 present various challenges and opportunities, however, one common challenge is finding the best position to deploy the UAVs. Despite the massive growth in the integration of UAVs in cellular networks, most of the literature still focuses on analytical solutions for positioning UAVs in mobile networks. For instance, [3] derive an expression for determining the optimal position of an UAV equipped with a radio module to maximize its coverage area considering energy constraints, while [12] find a closed-form equation for the UAV altitude which maximizes its coverage area. Also regarding coverage, [13] derive the average coverage probabilities for static UAVs and [14] determine analytically how to place multiple UAVs in an emergency scenario to reduce interference and maximize throughput. On the other hand, [15] propose and solve a convex optimization problem to find the best 3D position of multiple UAVs. Another solution, studied by [6], involves the utilization of UAVs with free space optics backhaul capabilities; their results show that higher backhaul data rates can be achieved. Lastly, one of the few solutions that somewhat leverages the adaptability and flexibility of ML is the work by [16] in which they determine the optimal number and position of multiple UAVs using particle swarm optimization.

As can be seen, most of the literature attempts to find the optimal placement of one or multiple UAVs in an analytical manner. Although these solutions have their merits, they often require unrealistic assumptions, such as knowledge of the number of users in the network, their positions or that users remain static. These assumptions are bold and might not be true in most real scenarios, making the proposed solutions ineffective to be used in practical situations. In contrast, in [4] we propose an intelligent solution, based on RL, to autonomously position multiple UAVs without previous knowledge of the environment (model-free). Table 10.1 presents a summary of the state-of-the-art literature with respect to positioning UAVs in the context of cellular networks.

Table 10.1 State-of-the-art UAV positioning solutions in cellular networks.

Reference	Type	Short Summary
Wang et al. [3]	A	UAV position to maximize coverage area considering energy constraints
Al-Hourani et al. [12]	A	Closed-form equation for optimal UAV altitude
Merwaday and Guvenc [14]	A	Optimal UAV position in an emergency scenario
Bor-Yaliniz and Yanikomeroglu [15]	A	Optimal positioning of multiple UAVs
Mozaffari et al. [13]	A	Outage and coverage probability of UAVs
Kalantari et al. [16]	A	Best UAV position considering backhaul constraints
Alzenad et al. [6]	A	Provision of aerial backhaul to the core network through UAVs
Kalantari et al. [17]	M	Positioning of multiple UAVs via Particle Swarm Optimization
Klaine et al. [4]	RL	Positioning of multiple UAVs via Q-learning

A – Analytical
M – Metaheuristic
RL – Reinforcement Learning

10.4 Reinforcement Learning

When the environment is dynamic and it is challenging to find a fixed or static solution, RL is a great candidate, as its algorithms are able to learn online from experiences and find optimal solutions [18]. Since we are dealing with a network that is constantly changing (for instance, users are moving, channel conditions and interference patterns are changing, user requirements can be very different, etc), we explore RL for positioning multiple UAVs in the context of an emergency network. For the reader's benefit, we have included in this section a small summary of how RL works.

In RL, a system, also known as an agent, learns by exploring the environment through actions and observing rewards. In this case, the agent is not told what actions to take, but instead it must discover which actions yield the best rewards by trying them [18].

One branch of RL consists of Value-Based algorithms, which rely on the determination of a Value Function or an Action Value Function. These functions are represented by tables which describe how good being in a state is – Value Function – or how good taking an action in a given state is – Action Value Function. To update these tables, the agent explores the environment taking actions and observing rewards, with the goal of maximizing the total cumulative reward [18]. This interactions are depicted in Figure 10.2. Next we provide a brief description of Q-learning, one of the most popular RL algorithms.

10.4.1 Q-Learning

In Q-learning, the agent finds itself in certain state s_t. Then it takes an action a_t which takes him to a new state s_{t+1} and observes a reward r_{t+1} from the environment. Next, it updates its Q-table entry for s_t and a_t according to what it already had, the observed reward r_{t+1} and the highest value on the table for all the actions at s_{t+1}. This can be summarized as [19]

$$Q(s_t, a_t) \leftarrow (1 - \lambda)Q(s_t, a_t) + \lambda[r_{t+1} + \phi \max_a Q(s_{t+1}, a)], \tag{10.1}$$

where λ is the learning rate, which represents how much the agent values past experiences, and ϕ is the discount factor, which represents how much the agent values future rewards.

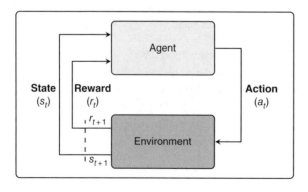

Figure 10.2 Block diagram of how RL works [1].

Q-learning is a useful tool, in particular because it has been shown to converge as long as all the states have the possibility of being visited infinitely many times given an infinite time to explore [20]. After *Q*-learning has converged, the agent can always choose the highest value of the *Q*-table to decide which actions to take, leading to high rewards, in what is known as exploitation. However, before converging, if the agent always decides to choose the highest value, it will barely explore the environment, and it might be stuck at local minima. This is known as the exploration-exploitation trade-off in RL [18].

Since in RL the agent implements a mapping from states to probabilities of selecting each possible action (known as the policy), and this policy changes over time as the agent traverses the environment, usually, it is preferable to explore more in the beginning, since the agent does not have much idea on which states and actions are good, while later on it is better to exploit. This can be achieved, for instance, by choosing a random action with probability ϵ. By making ϵ decay over time the ratio between exploitation and exploration changes as desired [18].

10.5 Simulations

In this section we explore the intelligent positioning strategy based on *Q*-learning. In the end it will become clear to the reader the strength of RL when deploying UAVs in mobile networks. The considered scenario is similar to the one explored in [4]; however, here we consider a few more users in total. Furthermore, all users are assumed to have higher mobility and require more throughput. Lastly, here we assume a smaller ratio of rescue users.

10.5.1 Urban Model

As defined by the International Telecommunication Union (ITU-R) in [21], any urban environment can be characterized by three parameters to account for the building topology. These parameters are:

- α, the ratio of build-up land area to the total land area;
- β, the average number of buildings per square kilometre;
- γ, scale parameter for the heights of the buildings, following a Rayleigh random distribution.

Similarly to [4, 22], we use these parameters and a Manhattan grid layout and consider a square area with side $L = 1$ km, as in Figure 10.3, with blocks of width W separated by S meters.

Considering W and S constant across all blocks, it is possible to find their values via [22]

$$W = 1000\sqrt{\frac{\alpha}{\beta}}, \tag{10.2}$$

and

$$S = \frac{1000}{\sqrt{\beta}} - W. \tag{10.3}$$

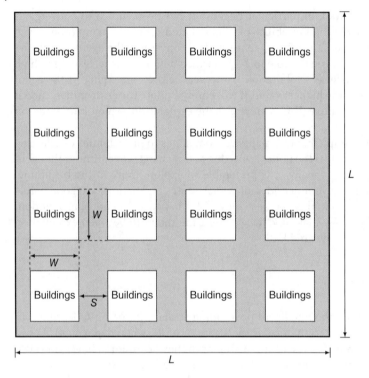

Figure 10.3 Manhattan grid urban layout [4].

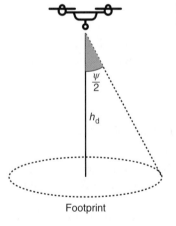

Figure 10.4 UAV flying at a height h_d, and with an antenna with an aperture angle of ψ. Source [4].

10.5.2 The UAVs

The UAVs are equipped with an antenna which has strong directivity, as seen in Figure 10.4, which means that the signal coming from its major lobe concentrates the majority of its power up to an aperture angle of ψ. This is important for the case of multiple UAVs providing connectivity with a shared spectrum to mitigate the interference between UAVs.

10.5.3 Path Loss

The path loss between UAVs and users (PL_d) follows the model presented in [22], wherein PL_d is split between a free-space path loss and an additional loss, depending on whether or not there is line of sight (LOS) between the UAV and the user:

$$\text{PL}_\text{d} = 20 \log_{10}\left(\frac{4\pi f_c d_\text{d}}{c}\right) + \xi, \tag{10.4}$$

where d_d is the distance between the UAVs and the users, c is the speed of light, f_c is the carrier frequency and ξ is the additional loss which assumes different values for LOS or non-line-of-sight (NLOS) links, 1 and 20 dB respectively [22].

Furthermore, a macro BS is connected to the core network. The macro BS has an out of band connection to the UAVs, thus connecting them to the backhaul. In addition, this macro BS can also serve to connect nearby users. The path loss between the macro BS and the users can be written using the Okumura–Hata urban model [23]

$$\text{PL}_\text{m} = 69.55 + 26.16 \cdot \log_{10}(f_c) - 13.82 \cdot \log_{10}(h_\text{B}) - C_\text{H} +$$
$$+ [44.9 - 6.55 \cdot \log_{10}(h_\text{B})]\log_{10}(d_\text{m}), \tag{10.5}$$

where d_m is the distance between a user and the macro BS, h_B is the height of the macro BS and C_H is the antenna height correction factor for urban scenarios, which is given by

$$C_\text{H} = 0.8 + (\log_{10}(f_c) - 0.7) \cdot h_\text{u} - 1.56 \cdot \log_{10}(f_c), \tag{10.6}$$

where h_u is the height of users' mobile devices.

10.5.4 Simulation Scenario

The simulation is performed in an area where there were previously one macro base station (BS) and several small cells. After a disaster has occurred, the original infrastructure was destroyed and only parts of the original backhaul connection remained available. To restore connectivity as fast as possible, a macro cell in a temporary fixed infrastructure (such as a truck-mounted BS) is deployed near the location of the destroyed macro BS, in order to provide connectivity to the core network as well as to connect nearby users. This connection may be either via an undamaged optical fibre from the previous network, if there is any left after the disaster, or via a wireless radio link (e.g. satellite, microwave, etc) in other situations. In addition, UAVs are also deployed to perform the role of the previous small cells, forwarding the traffic from their users to the macro BS to reach the core network.

Furthermore, two different types of users are considered: users from rescue teams and regular users. Depending on the type of a user, different characteristics are given in terms of mobility levels or throughput requirements. For example, regular users can be either static or move slowly, as they represent people who might be stuck under debris, trapped or even waiting for help on given locations. On the other hand, search and rescue teams move faster, as they are more equipped and prepared to deal with emergencies and might be inside rescue vehicles. In terms of throughput, the same logic applies: regular users just want to send simple messages to tell their relatives or friends that they are OK, which requires less throughput, whereas rescue teams require higher throughput as they might want to use video feeds, communicate with other team members extensively or even use more elaborate applications.

Before the disaster struck, the cellular network had hot spots of users, where it provided small cells for extended coverage. These hot spots are places where there is more concentration of people, and after the disaster occurs those people are likely to still be near the hot spot areas. For this reason, we have randomly distributed a third of the users across the entire $1km^2$ area, while the other two thirds are evenly assigned and randomly placed at the locations of the hot spots. Regarding frequency bands, we assume that both UAVs and macro cell share the same frequency band, meaning that UAVs and macro cell interfere with each other and a frequency reuse factor of 1 is considered. However, in order to mitigate the interference between UAVs, we consider that each UAV has a single antenna with an elevation and horizontal plane apertures of $\psi = 60°$, which is a good approximation of commercially available antennae [24]. Thus, UAVs have fixed radius of coverage, varying with their altitude, and users out of that radius perceive a very low signal coming from that UAV. Furthermore, we consider that both the truck-mounted BS and the UAVs have limited RAN resources and that they share a 10MHz bandwidth, which correspond to a capacity of 50 Resource Blocks (RBs), according to Long Term Evolution (LTE) parameters. Moreover, we assume an ideal backhaul on the truck-mounted BS [25, 26], while the UAVs have limited backhaul capacity. The parameters used in the simulation are summarized in Table 10.2.

10.5.5 Proposed RL Implementation

We propose using Q-learning to determine the best position of multiple UAVs to work as extension of the RAN in a deployment which has a macro BS. The goal is to find a position for each UAV, which minimizes the number of users in outage. Additionally, each UAV keeps its own independent Q-table, making this a multi-agent distributed solution. The Q-learning model for the proposed scheme is as follows:

- The affected area where users are spread is the **environment**.
- UAVs are the **agents**, which explore the environment and take the actions, observing the rewards.
- The **state** s_t is the three-dimensional position of a UAV at instant t. It can only assume discrete values, since the Q-table must store an entry for all possible states and actions.
- Likewise, **actions** can only assume discrete values and are modelled as the agent moving one step[1] in any direction (forward, backward, right, left, up and down) or not moving at all.
- Since the goal is to reduce the overall number of users in outage, the **reward** is total number of users served between the UAVs and the macro BS. This requires that the UAVs exchange this information among themselves and with the macro BS.
- The agents use an ϵ-greedy **policy**, wherein an agent selects the maximum value of its Q-table most of the time, but selects a random action with probability ϵ (which decays over time).

Due to the nature of a simulated environment, episodes are introduced. An episode can be described as one snapshot of the network, which corresponds to one round of Q-learning optimization followed by movement of users. During the Q-learning round,

---

1 The step size depends on which discrete values of state are allowed.

Table 10.2 Simulation Parameters [15].

Parameters	Value
Ratio of build-up to total land area, α	0.3 [21]
Average number of buildings, β	500 buildings/km^2 [21]
Scale parameter for building heights, γ	15m [21]
ξ LOS	1 dB [22]
ξ NLOS	20 dB [22]
Side of the square area, L	1km
UAVX-axis step	50m
UAVY-axis step	50m
UAVZ-axis step	100m
Minimum UAV height	200m
Maximum UAV height	1,000m
User$X$-axis step (low/high)	5/15m
User$Y$-axis step (low/high)	5/15m
Number of users, N_u	800 [26]
User height, h_u	1.5m
Ratio of rescue team users	10%
Number of hot spots	16
Number of UAVs	16
Ratio of users near hot spots	2/3 [26]
Macro BS Equivalent Irradiated Power	0 dBW [26]
Macro BS height, h_B	20m
UAV Equivalent Irradiated Power	−3 dBW [27]
UAV antenna directivity angle, ψ	60° [24]
RBs in Macro Cell	50 [26]
RBs per UAV	50 [26]
Macro cell backhaul capacity	100Gbps [26]
UAV backhaul capacity	37.5Mbps/UAV [26]
Bandwidth of one RB	180kHz [26]
Carrier frequency, f_c	1GHz
High SINR requirement	10dB
Low SINR requirement	3dB
Total number of episodes	50
Number of independent runs	30
Max iterations per episode, Max_it	1,000
Max iterations, same reward, $Max_\text{it,r}$	100
Min iterations per episode, $Min_\text{it,r}$	200
Learning Rate, λ	0.9
Discount Factor, ϕ	0.9

each UAV keeps its own Q-table and explores until one of the stopping criteria has been met. Once all UAVs have stopped, the users move and a new episode is started. The stopping criteria are as follows:

1. The UAV has moved for a maximum number of iterations Max_{it}.
2. The value of the reward has not improved in a certain number of iterations $Max_{it,r}$.
3. The UAV has used all its resource blocks and has explored for a minimum number of iterations Min_{it}.

To highlight how important it is to have an intelligent positioning system, we compare the intelligent approach using Q-learning with other static solutions. It is important to note that, for fairness, we cannot assume that the static positions have any knowledge about the users distribution. For this reason, the fixed solutions found in the literature and discussed in Section 10.3 cannot be used here for comparison. Therefore, we compare the intelligent approach with deploying the UAVs in the locations of the previous small cells, which existed prior to the disaster.

10.5.5.1 Simulation Results

Having an intelligent and moving strategy to find where to deploy the UAVs when considering an ECN is essential to provide wide coverage. As can be observed in Figure 10.5, the fixed solution cannot improve the coverage over time and, moreover, always yield worse performance when compared to Q-learning. Furthermore, the proposed RL solution is able to improve its performance significantly as time passes since the UAVs explore the environment and are able to find where are the users that need to be served. Moreover, the performance keeps improving as the users move, since the UAVs are always exploring and finding better positions to provide connectivity.

Albeit providing good coverage, the Q-learning-based solution is not able to improve the quality of experience (QoE) of users. As we can see on Figure 10.6, which shows

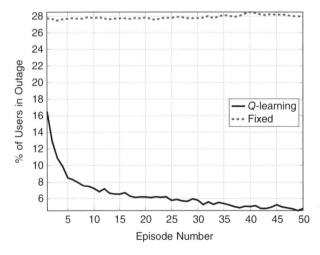

Figure 10.5 Users in outage for every episode, comparing positioning the UAVs in previous hot spot locations with the intelligent Q-learning solution on the ECN scenario.

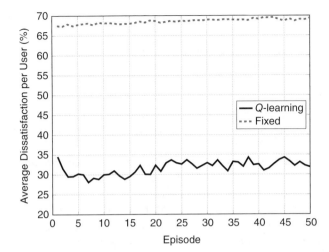

Figure 10.6 Average dissatisfaction of users on the ECN scenario, considering Q-learning and a fixed positioning strategy.

the average dissatisfaction of users, both strategies remain roughly constant in terms of this metric. Still, the Q-learning solution has the best performance. The reason Q-learning is not able to improve the QoE over time is because there is no information of the throughput dissatisfaction in the reward. Moreover, the primary objective is to get users connected (as is the priority on emergency situations), and devising a reward to improve the dissatisfaction might come at the cost of connecting fewer users. In this case, if we wish to improve the QoE, a more flexible – but at the same time, also intelligent – strategy has to be employed, which allows us to include other parameters in the model.

10.6 Conclusion

UAVs are expected to play a crucial role in next generation of mobile networks, enabling future applications such as ECNs, flying backhaul/fronthaul, caching on the clouds, pop-up networks, and many others that are unimaginable today. However, for that to be possible, an intelligent positioning strategy is essential, in order to guarantee the aerial network adaptability and autonomy.

It is clear from the simulations that using intelligent positioning algorithms based on RL is a viable strategy for that. However, more research is needed to design intelligent solutions which can improve multiple KPIs and generalize from past experiences at the same time. For instance, value function approximation [18] can be used to estimate the action value function based on features – thus including more information about the environment in the model – therefore, similar states would yield similar action values, providing the desired generalization from past experiences. This can be implemented, for instance, considering linear combinations of features or with more robust techniques such as Deep Q-network [28] or double Q-learning [29].

References

1 P. V. Klaine, M. A. Imran, O. Onireti, and R. D. Souza. A survey of machine learning techniques applied to self-organizing cellular networks. *IEEE Communications Surveys Tutorials*, 19(4):2392–2431, Fourthquarter 2017. doi: 10.1109/COMST.2017.2727878.

2 Mohammad Mozaffari, Walid Saad, Mehdi Bennis, Young-Han Nam, and Mérouane Debbah. A tutorial on UAVs for wireless networks: Applications, challenges, and open problems. *arXiv preprint arXiv:1803.00680*, 2018.

3 L. Wang, B. Hu, and S. Chen. Energy efficient placement of a drone base station for minimum required transmit power. *IEEE Wireless Communications Letters*, pages 1–1, 2018. ISSN 2162-2337. doi: 10.1109/LWC.2018.2808957.

4 Paulo V. Klaine, João P. B. Nadas, Richard D. Souza, and Muhammad A. Imran. Distributed drone base station positioning for emergency cellular networks using reinforcement learning. *Cognitive Computation*, May 2018. ISSN 1866-9964. doi: 10.1007/s12559-018-9559-8.

5 M. Chen, M. Mozaffari, W. Saad, C. Yin, M. Debbah, and C. S. Hong. Caching in the sky: Proactive deployment of cache-enabled unmanned aerial vehicles for optimized quality-of-experience. *IEEE Journal on Selected Areas in Communications*, 35(5):1046–1061, May 2017. ISSN 0733-8716. doi: 10.1109/JSAC.2017.2680898.

6 M. Alzenad, M. Z. Shakir, H. Yanikomeroglu, and M. S. Alouini. FSO-based vertical backhaul/fronthaul framework for 5G+ wireless networks. *IEEE Communications Magazine*, 56(1):218–224, Jan 2018. ISSN 0163-6804. doi: 10.1109/MCOM.2017.1600735.

7 Somen Nandi, Saigopal Thota, Avishek Nag, Sw Divyasukhananda, Partha Goswami, Ashwin Aravindakshan, Raymond Rodriguez, and Biswanath Mukherjee. Computing for rural empowerment: enabled by last-mile telecommunications. *IEEE Communications Magazine*, 54(6): 102–109, 2016.

8 Lina Xu, Rem Collier, and Gregory MP O'Hare. A survey of clustering techniques in WSNs and consideration of the challenges of applying such to 5G Iot scenarios. *IEEE Internet of Things Journal*, 4(5):1229–1249, 2017.

9 Joao Pedro Battistella Nadas, Richard Demo Souza, Marcelo Eduardo Pellenz, Glauber Brante, and Sergio Michelotto Braga. Energy efficient beacon based synchronization for alarm driven wireless sensor networks. *IEEE Signal Processing Letters*, 23(3):336–340, 2016.

10 Fauzun A Asuhaimi, João PB Nadas, and Muhammad A Imran. Delay-optimal mode selection in device-to-device communications for smart grid. In *Smart Grid Communications (SmartGridComm), 2017 IEEE International Conference on*, pages 26–31. IEEE, 2017.

11 Afif Osseiran, Federico Boccardi, Volker Braun, Katsutoshi Kusume, Patrick Marsch, Michal Maternia, Olav Queseth, Malte Schellmann, Hans Schotten, Hidekazu Taoka, et al. Scenarios for 5g mobile and wireless communications: the vision of the metis project. *IEEE Communications Magazine*, 52(5):26–35, 2014.

12 A. Al-Hourani, S. Kandeepan, and S. Lardner. Optimal lap altitude for maximum coverage. *IEEE Wireless Communications Letters*, 3(6): 569–572, Dec 2014b. ISSN 2162-2337. doi: 10.1109/LWC.2014.2342736.

13 M. Mozaffari, W. Saad, M. Bennis, and M. Debbah. Unmanned aerial vehicle with underlaid device-to-device communications: Performance and tradeoffs. *IEEE Transactions on Wireless Communications*, 15(6): 3949–3963, June 2016. ISSN 1536-1276. doi: 10.1109/TWC.2016.2531652.

14 Arvind Merwaday and Ismail Guvenc. UAV assisted heterogeneous networks for public safety communications. In *Wireless Communications and Networking Conference Workshops (WCNCW), 2015 IEEE*, pages 329–334. IEEE, 2015.

15 I. Bor-Yaliniz and H. Yanikomeroglu. The new frontier in RAN heterogeneity: Multi-tier drone-cells. *IEEE Communications Magazine*, 54(11):48–55, November 2016. ISSN 0163-6804. doi: 10.1109/MCOM.2016.1600178CM.

16 E. Kalantari, M. Z. Shakir, H. Yanikomeroglu, and A. Yongacoglu. Backhaul-aware robust 3D drone placement in 5G+ wireless networks. In *2017 IEEE International Conference on Communications Workshops (ICC Workshops)*, pages 109–114, May 2017. doi: 10.1109/ICCW.2017.7962642.

17 E. Kalantari, H. Yanikomeroglu, and A. Yongacoglu. On the number and 3D placement of drone base stations in wireless cellular networks. In *2016 IEEE 84th Vehicular Technology Conference (VTC-Fall)*, pages 1–6, Sept 2016. doi: 10.1109/VTC-Fall.2016.7881122.

18 Richard S Sutton and Andrew G Barto. *Reinforcement learning: An introduction*, volume 1. MIT press Cambridge, 1998.

19 Christopher J. C. H. Watkins and Peter Dayan. Q-learning. *Machine Learning*, 8(3):279–292, May 1992. ISSN 1573-0565. doi: 10.1007/BF00992698. URL https://doi.org/10.1007/BF00992698.

20 Francisco S Melo. Convergence of Q-learning: A simple proof. *Institute Of Systems and Robotics, Tech. Rep*, pages 1–4, 2001.

21 ITU-R. *Rec. P. 1410-2 Propagation Data and Prediction Methods for The Design of Terrestrial Broadband Millimetric Aadio Access Systems*, 2003.

22 A. Al-Hourani, S. Kandeepan, and A. Jamalipour. Modeling air-to-ground path loss for low altitude platforms in urban environments. In *2014 IEEE Global Communications Conference*, pages 2898–2904, Dec 2014a. doi: 10.1109/GLOCOM.2014.7037248.

23 Theodore S Rappaport et al. *Wireless communications: principles and practice*, volume 2. prentice hall PTR New Jersey, 1996.

24 Fullband fbxpmimo outdoor 4G antenna – cross polarised 3G/4G antenna. http://www.fullband.co.uk/product/fbxpmimo/, 2017. Accessed: 2017-11-25.

25 M. Jaber, M. Imran, R. Tafazolli, and A. Tukmanov. An adaptive backhaul-aware cell range extension approach. In *Proc. IEEE International Conference on Communication Workshop (ICCW)*, pages 74–79, June 2015. doi: 10.1109/ICCW.2015.7247158.

26 M. Jaber, M. A. Imran, R. Tafazolli, and A. Tukmanov. A distributed SON-based user-centric backhaul provisioning scheme. *IEEE Access*, 4: 2314–2330, 2016. ISSN 2169-3536. doi: 10.1109/ACCESS.2016.2566958.

27 Mohammad Mahdi Azari, Fernando Rosas, and Sofie Pollin. Reshaping cellular networks for the sky: The major factors and feasibility. *arXiv preprint arXiv:1710.11404*, 2017.

28 Volodymyr Mnih, Koray Kavukcuoglu, David Silver, Andrei A Rusu, Joel Veness, Marc G Bellemare, Alex Graves, Martin Riedmiller, Andreas K Fidjeland, Georg

Ostrovski, et al. Human-level control through deep reinforcement learning. *Nature*, 518(7540):529, 2015.

29 Hado Van Hasselt, Arthur Guez, and David Silver. Deep reinforcement learning with double Q-learning. In *AAAI*, volume 2, page 5. Phoenix, AZ, 2016.

11

Integrating Public Safety Networks to 5G: Applications and Standards

Usman Raza, Muhammad Usman, Muhammad Rizwan Asghar, Imran Shafique Ansari and Fabrizio Granelli

11.1 Introduction

Reliable and robust communication networks are fundamental for Public Safety (PS) agencies to provide services for Public Protection and Disaster Relief (PPDR). Historically, Public Safety Networks (PSNs) and commercial mobile broadband evolved in isolation, pursuing divergent goals. PSNs transitioned from analog systems of the 20th century to today's digital narrowband systems, deployment of which started back in the 1990s. Their main focus was *mission-critical* voice services that enable communication between emergency first responders in one-on-one fashion or within a group with the help of deployed infrastructure or even without it. However, their narrowband nature does not permit high data rate multimedia services such as video that is instrumental for providing effective PPDR services.

Commercial broadband networks evolved in parallel but experienced a spectacular technological revolution, mostly driven by high profitability and large market size due to a wider public audience. As a result, techniques for efficient radio resource utilization over a broadband spectrum made them well-suited for high data rate services including voice, video, etc. However, unlike narrowband PSNs, these technologies provide *non-mission-critical* and often only one-to-one communication. The support for group communications remained very basic, meeting far less stringent constraints than those required for traditional PSNs.

A brief history of transition of PSNs from the analog system to future 5G networks is summarized in Table 11.1. To conduct their day-to-day operations, PS agencies combined complementary strengths of mission-critical PSNs and more pervasive commercial cellular networks. However, the lack of a single communication system in the field operations led to both operational and technological inefficiencies such as slower response time and sub-optimal spectrum usage. To overcome these limitations, different Standards Developing Organizations (SDOs) have made strides recently in proposing standards to support the convergence of both kinds of networks. The latest major advancement in this area is done by the 3rd Generation Partnership Project (3GPP) through the ratification of 5G.

Enabling 5G Communication Systems to Support Vertical Industries, First Edition.
Edited by Muhammad Ali Imran, Yusuf Abdulrahman Sambo and Qammer H. Abbasi.

Table 11.1 A brief history and evolution of Public Safety Networks (PSNs).

Year	Event
1923	First two-way radio developed
1939	First Walkie-Talkie designed
1941	First two-way radio system by Motorola deployed in Philadelphia, Pennsylvania, USA
1976	First group communications in PSN
1988	Initiative to migrate analog to digital PSNs
1989	First digital PSN, APCO Project 25, came into existence in USA
1995	TETRA came into existence in Europe
2005	DMR standardized by ETSI
2013	FirstNet LTE deployment started
2015	Support ProSe services and group communications for PS applications in 3GPP release 12
2016	Support for mission-critical push to talk for PS applications in 3GPP release 13
2017	Support for multimedia broadcast supplement for public warning system in 3GPP release 14.

In foreseeable future, PS agencies are expected to adopt broadband PSNs, while legacy narrowband networks will still coexist for a smoother transition during infrastructural and operational changes. This chapter summarizes standardization efforts made to support an inevitable convergence of PS communication networks across multiple regions. Further, we anticipate that this convergence will not be limited to narrowband and broadband technologies but will also be influenced by recent trends on how people access information today. Over the past few decades, information access methods have extended well beyond basic telephony. An ongoing wave of revolution in emerging technologies like massive Internet-of-Things (IoT), crowdsourcing, Online Social Networks (OSNs), Wireless Sensor Networks (WSNs), and many others combined with industry's push towards transition of communication systems to All-IP-based networks will change the landscape of future PSNs. All these trends are bound to blend into impending narrowband-broadband convergence for PSNs.

We envision that future PSNs will be an inclusive system that will glue different broadband and narrowband technologies along with other emerging technologies mentioned above. In this grand scheme, technology will enable the general public to get more involved in delivering their collective responsibility towards PS. Moreover, this chapter highlights different challenges to achieve this vision with some future directions. The requirements, challenges, and opportunities concerning PS communications are discussed in [1].

The rest of the chapter is organized as follows: In Section 11.2, we discuss possible PS scenarios in future 5G networks. Section 11.3 describes the standardization process of integrating legacy PSNs to commercial cellular networks. In Section 11.4, we present potential challenges and enabling technologies for this integration. Finally, Section 11.5 concludes this chapter.

11.2 Public Safety Scenarios

3GPP presents different use cases and scenarios for PS applications, mainly based on the coverage area of cellular networks. These scenarios are presented to support National Security and Public Safety (NSPS) services using Device-to-Device (D2D) communications in next-generation cellular networks [2, 3].

These scenarios can be categorized into three distinct types, which are elaborated in Figure 11.1. It is important to note that the scenarios presented in this section are for D2D communications only. For detailed use-cases and scenarios of PSNs in 5G, one may look into the work presented in [1].

11.2.1 In-Coverage Scenario

In this scenario, all User Equipments (UEs) lie in the coverage area of the cellular network. The network operator is fully responsible to control all functionalities of D2D communications, such as device discovery, device configuration, identity authentication, resource allocation, connection establishment and termination, access control, and security management. In this scenario, the D2D links are established in the same licensed spectrum on which UEs are connected to the Base Station (BS). In other words, UE-to-UE communications and UE-to-BS communications share the same licensed spectrum.

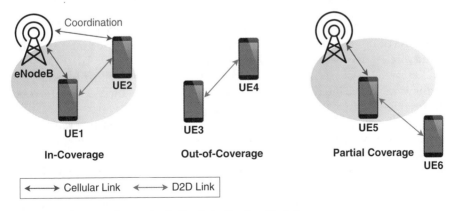

Figure 11.1 Supported scenarios for Proximity Services (ProSe).

11.2.2 Out-of-Coverage Scenario

In this scenario, all UEs are located outside the coverage area of the cellular network. The devices can establish D2D links without any assistance from the infrastructure. This use-case is the practical application of the disaster scenarios, where the cellular infrastructure is completely or partially damaged and a part of the network is offline. For example, in Figure 11.1, UE3 and UE4 can establish D2D links and start their communication without any support from the infrastructure. In this scenario, the UEs are responsible to control most of the functionalities of D2D communications, such as device discovery, device configuration, identity authentication, connection establishment and termination, access control, and security management. In 3GPP standardization procedures, communication in the out-of-coverage scenario is available for PS UEs only, while commercial UEs cannot avail themselves of this service.

11.2.3 Partial-Coverage Scenario

In this scenario, some UEs are located outside the coverage area of the cellular network. The UEs at the edge of the coverage area relay the information of out-of-coverage UEs to the base station or core network. This helps to extend the coverage area of the network at the edge. For example, in Figure 11.1, UE5 acts as a relay node to extend the coverage of the cellular network to UE6. In this scenario, the network operator is fully responsible to control all functions of D2D communications, such as device discovery, device configuration, identity authentication, resource allocation, connection establishment and termination, access control, and security management, for both UE-to-UE and UE-to-BS communications.

11.3 Standardization Efforts

This section highlights the efforts to standardize the inter-operable broadband technology for PSNs. It is important to note that 3GPP is the main contributor to the standardization process of broadband PSNs; however, 3GPP is not the only contributor. In fact, some other SDOs are also playing an important role to make a smooth transition from narrowband to inter-operable broadband PSNs.

In 2012, the First Responder Network Authority (FirstNet) [4] was created in the United States (US) to build, operate, and maintain the first nationwide inter-operable broadband wireless network for PS. The FirstNet, in support with Public Safety Communication Research (PSCR), is coordinating with SDOs for setting standards to advance PS communications and interoperability as shown in Figure 11.2. These SDOs are working towards different aspects of broadband PSNs.

3GPP is mainly responsible for standardizing the broadband supplement of all features that we have in narrowband technology (both for voice and video), e.g., Mission-Critical Push-To-Talk (MCPTT), group communications, and Public Warning System (PWS). Similarly, Alliance for Telecommunication Industry Solution (ATIS) is mainly responsible for Next Generation 911 (NG911) and all-IP-based solutions for interoperability. On the other hand, Groupe Speciale Mobile Association (GSMA) is providing support for their Voice over LTE (VoLTE) initiative. Moreover, Open Mobile Alliance (OMA)

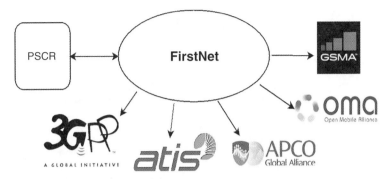

Figure 11.2 FirstNet PS standardization process.

is working on the standardization of Push to Communicate for Public Safety (PCPS). Lastly, The Association of Public-Safety Communications Officials – International, Inc. (APCO International) is committed to making the best use of emerging technologies for narrowband Land Mobile Radio (LMR) equipment and systems.

In what follows, we will briefly discuss the role of each SDO towards standardization of broadband PSN.

11.3.1 3rd Generation Partnership Project

Since the interest of PS community in broadband PSNs, 3GPP has been actively participating to ensure a smooth transition of narrowband PSNs to broadband networks (such as 5G). In particular, 3GPP[1] takes the requirements and technical inputs from the standards committees of the narrowband PSN (such as Terrestrial Trunked Radio (TETRA) and Project 25 (P25)) to enhance the LTE technology to support the required features [5]. Technically, the Technical Specification Groups (TSGs) working on Radio Access Network (RAN) are responsible for enhancing the basic functionalities of the access network to support direct D2D communications. However, for the overall applications on top, i.e., voice communications (known as push-to-talk) and group communications, a separate standardization group has been created. This group is called "Service and System Aspects Working Group 6 (SA WG6)", shortened to SA6.

Since LTE is a cellular technology and most of its specifications were completed in 3GPP release 8, we start from release 8 highlighting PS features. However, a summary of all supported features of PSNs in 3GPP is presented in Figure 11.3.

11.3.1.1 Release 8
3GPP release 8, completed in December 2008, provided support for LTE band 14, a 20MHz spectrum in 700MHz band, dedicated to PS both in the US and Canada. In addition, release 8 provided an ability to set up a mobile data connection from a device to packet network [6]. These are the very basic features of broadband PSNs.

11.3.1.2 Release 9
3GPP release 9, completed in December 2009, included several enhancements related to PS [6]. It provided support for Location-Based Services (LBS) and positioning, which

---

1 http://www.3gpp.org

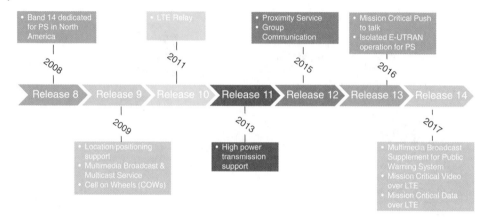

Figure 11.3 Evolution of PS features in 3GPP.

are critical for first responders to track each other during a mission. The most important PS feature in release 9 was the support of Multimedia Broadcast and Multicast Service (MBMS). It is an enabling technology for group communications and video broadcasting in PSNs. More importantly, Release 9 also provided an emergency calling support, for instance 911, which is critical to ensure PS. In addition, for PS scenarios, release 9 provided a support for the mobile base station, *a.k.a.* Cell on Wheels (COW) in the PS community.

11.3.1.3   Release 10
3GPP release 10 (LTE-Advanced), completed in March 2011, provided relay support for LTE. This allows a base station, for example mounted on a fire vehicle, to relay information from a firefighter in a basement back to the network [7].

11.3.1.4   Release 11
3GPP release 11, completed in March 2013, included the support for high power devices in band 14. Generally, the power for normal cellular devices is limited to 200mW, while in this release the power for PS devices was increased to 1.25 Watts. This significantly reduced the network deployment cost, improving coverage area of LTE network for PS scenarios [8].

11.3.1.5   Release 12
3GPP release 12 was one of most important releases, and incorporated two very important features of PS communications, *i.e.*, Proximity Services (ProSe) and Group Communications. ProSe, *a.k.a.* D2D communications, is an enabler of direct communication between PS devices, while group communication allows PS users to operate in groups, especially when they are on mission [9]. This release was completed in March 2015.

In what follows, we will briefly describe the main features of ProSe and group communications in LTE release 12.

Proximity Services (ProSe): The proximity services in LTE are about discovering mobile devices in physical proximity and enabling communication among them with or without supervision from the LTE network. These services are highly critical for PS

officials and first responders. In release 12, these services are sometimes described as
either "Direct Mode" or "off-network communication". The system enablers for ProSe
services include direct discovery and direct communication, Evolved Packet Core
(EPC)-level ProSe discovery, and EPC support for WLAN-based direct discovery and
communications.

The scenarios that are supported by direct discovery and direct communications
are listed in Table 11.2 and shown in Figure 11.1. In particular, new network elements
are added and new interfaces are defined to support ProSe in LTE. A simplified block
diagram of overall network architecture, including ProSe functionalities, is shown
in Figure 11.4 [10]. From a direct communication point of view, the most important
interface is PC5 interface between two UEs and PC3 interface between the UE and
new network node, ProSe function. The information exchange through PC3 interface is

Table 11.2 Supported ProSe functions in 3GPP release 12 to enable D2D
communications in public safety and non-public safety applications.

Scenarios	Within Network Coverage	Outside Network Coverage	Partial Network Coverage
Supported Applications		*Supported ProSe Functions*	
Non-Public Safety	Discovery	-	-
PS	Discovery Communication	Communication	Communication

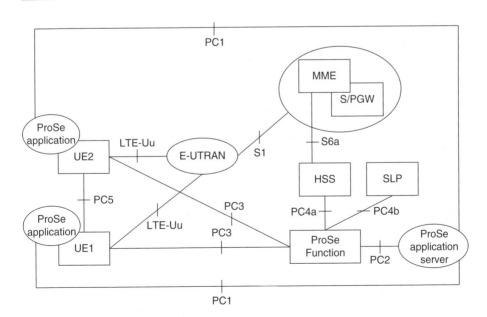

Figure 11.4 D2D ProSe network architecture [10].

mainly based on HTTP, which is used as a transport protocol. The ProSe messages are embedded in the body of HTTP Request and Response messages.

For out-of-coverage scenarios, the most relevant parameters related to PS can be stored in the UE itself, either hardcoded into the UE or in the Universal Integrated Circuit Card (UICC) or Subscriber Identity Module (SIM card). The ProSe function is responsible for the authorization and provision of radio and security parameters for direct communication.

Group Communication: This is one of the most important and frequently used features of PSNs when communicating with multiple users. The group communication is dependent on MBMS services introduced in LTE release 9. In release 9, there was no mechanism to measure the signal quality at the UE needed for providing better tools to monitor and adjust the MBMS operational parameters. This is included in release 12 by adding new measurements (e.g., signal strength, signal-to-noise ratio, and error rate) in the radio layer Multicast Broadcast Single Frequency Networks (MBSFN) metric. The MBMS access is made available by the creation of MB2 interface between Group Communication Service Application Server (GCS-AS) and Broadcast Multicast Service Center (BMSC). The MBMS radio resources in a cell are semi-statically allocated according to low activity period in order not to overprovision them uselessly. But if there is an incident and several PS groups in that cell suddenly need to communicate over those MBMS resources, this results in a severe overload of these resources. The group communication work item in release 12 provides mechanisms to cope with this overload situation.

11.3.1.6 Release 13

3GPP release 13 was completed in March 2016. In release 13, several PS features initiated in release 12 were continued for further enhancement [11]. For instance, LTE D2D communications and discovery were further enhanced to meet the PS requirements for in-network coverage (inter- and intra-cell), partial network coverage, and outside network coverage scenarios. The non-PS discovery is still limited to in-network coverage only. There are further enhancements to D2D communications to enable the extension of the network coverage using Layer 3 based UE-to-Network relay for partial and outside network coverage scenarios, considering applicability to voice and video. The most important PS features in release 13 are summarized here.

Mission-Critical Push-to-talk over LTE (MCPTT): MCPTT is the key enabler for many PS features, such as person-to-person calls, group calls, group management, user management, security, operation in off-network mode, operation in relay-to-network mode, and many other related features. MCPTT uses ProSe services to enable direct communication between devices, both in and out of LTE network coverage. Current standardization efforts include the development of application layer architecture for MCPTT, defining basic architecture functionalities across applications, and identifying reusable components in existing TETRA, Critical Communications Evolution (TCCE), and OMA PCPS specifications. A PTT application server is proposed as a part of GCS-AS, where GCS is a generic function for voice and data.

Related to MBMS, two areas are specifically enhanced in release 13. One is service continuity and the second is the independence of application from knowing the service areas defined in the network.

Isolated E-UTRAN Operation for PS: To provide voice, video, and data services to PS officials working in an area where backhaul connection is cut off during mission-critical operation, the isolated E-UTRAN operation for PS is proposed in release 13. The PS officials may deploy dedicated eNodeB(s) for nearby PS UEs beyond what is provided by UE-to-UE direct communication mode. This requires enabling separate and autonomous security mechanisms for isolated E-UTRANs. In release 13, the isolated E-UTRAN operation for PS comprises the following features: operation with no or limited connection to EPC, multiple isolated eNodeBs with interconnection between them, and the support of local operations like group communication.

11.3.1.7 Release 14
3GPP release 14 improves the following features of PS communications [11, 12].

Multimedia Broadcast Supplement for Public Warning System (PWS): The purpose of this feature is to enable effective delivery (and broadcasting) of extensive multimedia public warning contents in PWS and investigating mechanisms to enable users to receive and view this content.

Mission Critical Video over LTE (FS_MCVideo): This includes enabling broadband features like video group communication (to enable see what I see), video broadcast, and video consolation to allow a PS UE to view a video with video recorder functionalities and manage several video streams in single operation. To accomplish this, the FC_MCVideo aims at reusing existing standards when possible and justified, including the MCPTT framework (e.g. call commencement, call termination, security, and user management). The FC_MCVideo requirements include the use of ProSe, security, confidentiality, interaction with MCPTT voice services for same functions like group management, group communication involving several media types (e.g., a user with voice commenting a video), and fall-back mode in case of scarce resources (e.g., possibility of transmitting a sound or commentary associated with a video).

Mission Critical Data over LTE (FS_MCDATA): Some of the requirements for this are inherited from different organizations like The Critical Communications Association (TCCA), APCO, OMA (Push-to-talk over Cellular (PoC) version 2.1 and PCPS), and The European Telecommunications Standards Institute (ETSI) TETRA and Critical Communications Evolution (TCCE) user requirements. Most potential requirements are almost the same as for FC_MCVideo.

11.3.1.8 Release 15
3GPP release 15 provides a further enhanced MCPTT group call procedure and provides support to integrate legacy PSNs to cellular networks [11]. In fact, release 15 includes the first of the two phases of the 5G standard, while release 16 will mainly encompass the second phase of 5G. It is important to note that release 15 mainly focuses on implementing the policy-based control of the network that makes it easier to integrate the PSNs by defining the policies regarding them, e.g. ensuring higher Quality-of-Service (QoS) and reliability etc.

In particular, release 15 works towards inter-working of legacy PSNs and 3GPP-defined mission-critical system. The release focuses on standardization of the interconnections between 3GPP-defined mission-critical systems. Moreover, this release partially covers the mission-critical requirements of the railway and maritime industries. The development of application program interfaces for MBMS is also being considered in 3GPP release 15. The details of these mission critical services are publicly available[2].

11.3.2 Open Mobile Alliance

Open Mobile Alliance (OMA)[3] is an industry forum that develops open specifications for enablers of new mobile services. The main objective for these specifications is to enable interoperability across different networks, regions, devices, and managing entities. OMA Communications Working Group (OMA COM WG) has defined multiple broadband Push to Talk (PTT) standards for commercial cellular networks and more recently for PSNs. We now briefly highlight other broadband PTT standards defined by OMA in the past [13].

11.3.2.1 PTT over Cellular

PTT over Cellular (PoC) is OMA's first broadband PTT standard for commercial cellular networks that has less stringent constraints on performance than PSNs Alliance [14]. This two-way radio standard provides support for immediate one-to-one and one-to-many communication. PoC is based on IMS and thus can be implemented on top of both 2G and 3G mobile systems.

After its first release in 2008 (v1.0), the standard was revised multiple times in 2009 (v1.04), and twice in 2011 (v2.0, v2.1) to incorporate new features. These features include support for automatic and manual answering modes, transmission of multimedia other than voice (such as text, files, videos, and images), and interconnection with external PTT networks. Although PoC is a commercial cellular standard, its latest version (v2.1) took initial steps towards providing mission-critical features desired for PS community such as the ability to communicate in ad-hoc mode, a multicast group or a pre-defined group. Furthermore, it can prioritize communications, allows pre-emption, and supports dispatcher functions.

11.3.2.2 Push to Communicate for Public Safety (PCPS)

PCPS is a newer standard from OMA to support both PS and commercial communications. It consolidates all features from PoC and extends them to work with LTE advanced (i.e., 3GPP Release 12). In 2015, OMA licensed PCPS V1.0 specifications to 3GPP for using it in defining MCPTT functionalities within the LTE standard.

With its close ties to other SDOs like 3GPP, OMA actively participated in defining MCPTT, a broadband PTT standard finalized in Release 13 of 3GPP that is discussed later in this section.

11.3.3 Alliance for Telecommunication Industry Solutions

Alliance for Telecommunication Industry Solutions (ATIS)[4] is one of the organizational partners of 3GPP that develops standards and solutions for Information and Communications Technology (ICT) industry. ATIS is working to support the transition of PS

---

2 http://www.3gpp.org/news-events/3gpp-news/1875-mc_services
3 https://www.omaspecworks.org
4 https://www.atis.org

applications from legacy circuit-switch networks to all-IP networks [15]. It realizes that several industrial sectors such as energy and utility, building alarms systems, public transportation, and emergency management services connect to PS agencies using legacy and non-standardized communication networks. These systems are capable of transporting voice, collecting sensed values, and applying supervisory control. However, their proprietary and application-specific nature makes them expensive and their horizontal integration a real challenge. The transition to All-IP is considered to protect critical infrastructure more reliably, assure interoperability across multiple applications, and reduce response time for handling emergencies in natural or man-made disasters.

In what follows, we highlight the targeted sectors of ATIS for all-IP transitions.

11.3.3.1 Energy and Utility Sector

In this PS sector, two communication solutions were proposed to replace PSTN copper wires. The first solution is based on an IP private line that can provide carrier-grade performance. The second is a wireless solution that deploys private wireless network across the entire grid. It is designed for network monitoring and system backup.

11.3.3.2 Building Alarm Systems

Police, fire brigade, and private security services usually respond to alarms triggered by private and public buildings and protected places such as airports and military installation to prevent loss of human life or property. Many of these alarms are still connected to the first responders using legacy circuit switched system. The first solution to migrate to IP is provided by a module that can transmit alarms to first responders and can continuously monitor IP connections with the server as well. The second is a Machine-to-Machine-based solution geared with a secure cellular (3G/4G) connectivity capable of reporting alarms, device status, and diagnostic information.

11.3.3.3 PS Communications with Emergency Centers

The evolution of PS wireless communications is already underway, as highlighted by standardization activities of 3GPP and OMA. In parallel to this, the integration of PS communications with Emergency Management Centres needs serious attention. ATIS has some roles in proposing the following solutions as a milestone towards migration to next-generation emergency services.

- Multiple broadband solutions for communication between state and federal emergency management agencies.
- An IP-based system that integrates with an emergency centre and radio dispatch via IP networks.
- Multiple solutions for NG911, supporting migration of analog to IP-based communications.
- Secure solution for call delivery using soft-switch routing, enhanced call handling, and secure IP networks.
- Voice over IP (VoIP)-based solution for communication with existing 911.

11.3.3.4 Smart City Solutions

PS is not the basis for smart city design and initial implementation, but there are opportunities for leveraging these deployments to support PS. In addition to working on mentioned PS sectors, ATIS, in collaboration with 3GPP, has initiated two standardization programs that will contribute to the long-term roadmap for PS applications: 5G-evolution and cybersecurity. These programs include solutions from different technology innovations like IoT, cloud applications, and Software-Defined Networking (SDN). The 5G-evolution program includes enhancements to redundancy and resiliency, security across networks, and capacity management during planned events and emergency response situations. Cybersecurity is accessing cutting-edge technologies that can offer innovative cybersecurity solutions in the future. These technologies include network virtualization, SDN, cloud architectures, and enhanced end-user control.

11.3.4   APCO Global Alliance

Association of Public-Safety Communications Officials (APCO)[5] is an international organization of PS communication professionals that establishes standards, solutions, and guidelines for interoperable, narrowband LMR/PMR systems in the areas of cybersecurity, NG911, and the telematics [16].

As part of their standardization efforts, APCO developed guidelines for the prevention of various cyber attacks on the IP-based PSNs and communications. This includes the prevention of Denial-of-Service (DoS) attacks, swatting/spoofing, and network intrusion attacks. Swatting/spoofing is the imitating of the caller's ID to trick the emergency services to dispatch personnel to a fake incident.

Another contribution made by APCO is the development of standards and guidelines in the area of vehicle telematics, which includes the handling of Advanced Automatic Crash Notification (AACN) calls from Telematics Service Providers (TSPs). The standard allows each agency to define their own policies for AACN call handling and dispatch.

APCO has also dedicated its efforts to develop standards for new forms of communication to 911 such as text-to-911 and mobile apps. The text-to-911 is an interim solution, which processes text messages to 911 via carrier native SMS. The solution does not depend on vendor-specific applications implemented in individual PS Answering Points (PSAPs). The solution provides direct access to 911 PSAPs for individuals having speech disabilities. Moreover, APCO is working to develop guidelines for the development of mobile applications, which can securely communicate with 911 centres. These mobile apps will use the public Internet, which can be a security threat to the NG911 network. The standard should ensure the security and reliability of NG911 centres.

In short, 911 is gradually transitioning to IP-based networks, which will enable the use of text, videos, and mobile apps with better security and reliability and APCO is playing its part to effectively integrate these new forms of communications to next-generation 911.

11.3.5   Groupe Speciale Mobile Association (GSMA)

GSMA[6], an association of mobile operators, supports the standardization of GSM system. Its VoLTE solution was fully standardized in 2000 [17] to provide a single digital

---

5 https://www.apcointl.org/about-apco/global
6 https://www.gsma.com

packet service capable of transferring both data as well as voice over LTE networks using IP. VoLTE is built on top of the IP Multimedia Subsystem (IMS) [17], a standard from 3GPP for providing multimedia services over cellular networks.

VoLTE provides a single homogeneous service for transferring both data and voice using IP, which eliminates the dependence on 2G/3G including the circuit-switched network. As a result, VoLTE provides features such as better spectrum utilization, higher voice quality, and better security. All these features make it an attractive option for PSNs as noted in [17] and industry is pushing it to become part of future PSNs. In addition, VoLTE provides wireline-quality voice services to LTE users, which is essential for mission-critical applications in PSNs. Nevertheless, its current specifications do not incorporate important PS features such as group communication, push-to-talk, and delivery of video among group Câmara & [18], an important gap that is bridged by the features introduced by 3GPP in releases 12–13, as discussed earlier.

Apart from standardization efforts from different SDOs, there is a need to develop a partnership across the industry to facilitate a transition pathway for critical applications. Although the manufacturers, network operators, emergency management agencies, and SDOs are providing different solutions, it is equally important to take a longer-term view of how this network evolution will present new opportunities and benefits that could be realized by the PS sector. It is important to note that in the PS sector, essential attributes such as reliability, resilience, latency, diversity, etc. will not be bounded by the performance characteristics of a single network (e.g., access or core); rather, the IP-enabled networks and devices will provide additional capabilities and media alternatives that, if designed properly, can provide a next-generation of PS applications.

11.4   Future Challenges and Enabling Technologies

The future PSN will be an inclusive system that will include different technologies, such as legacy professional mobile radio (Private Mobile Radio (PMR) in the United Kingdom (UK) and LMR in North America), massive IoT, commercial cellular networks, WSNs, Machine-to-Machine (M2M) communications, D2D communications, ad-hoc networks, satellite networks, aerial eNodeBs, OSNs, big data, and many others. The legacy PSNs will coexist with LTE for quite a long time until the new network is 100% flaw-proof. This complete system will be a part of 5G networks, where SDN will play a vital role in managing and operating this heterogeneous network in terms of technology both in access and core. All-IP-based solutions in access and NG911 in core network will help to make this complex informative system interoperable and integrated.

Massive IoT will be a part of the network, which will help to report critical incidents to NG911 without any human involvement. LBS will help to locate the site and the magnitude of the incident. D2D communications will keep the necessary communication until drones equipped with aerial eNodeB reach the site. The first responders will be able to communicate with each other and central station using D2D communication, aerial eNodeBs, and satellite communications.

Social media will be another tool to report incidents to NG911. Cloud computing and big data analytics tools will play a critical role in PS agencies' ability to detect and respond to threats.

The complete ecosystem of PSNs will enable real-time access to the information, where content-rich database lookups and remote analytics will lead to greater productivity and reduced costs. The system will enable broadband access to the PS officials to keep them connected anywhere and anytime. Mission critical data over LTE will give first responders capabilities to transmit a massive amount of data to and from command centres and/or directly among each other.

The ecosystem will greatly increase the utility and offer access to a whole array of advanced multimedia applications like bi-directional vehicular video, location-aware real-time services, mobile office, in-field productivity, multimedia command and control, dynamic mapping, weather and traffic flows, content-rich lookups to complex databases, and many others.

11.4.1 Future Challenges

PPDR organizations include law enforcement agencies, emergency management agencies, fire departments, Emergency Medical Services (EMS), rescue squads, and many others. These organizations make the use of the ICT sector for their day-to-day communications and rescue operations. The ICT sector for PS communication includes legacy PSNs (i.e. PMR/LMR) systems, broadband PSNs systems, massive IoT, commercial cellular networks, WSNs, OSNs, big data, and many others. Incorporating these diverse ICT technologies into PSNs poses different challenges. In the following section, we discuss the possible challenges and enabling technologies for future PSNs.

11.4.1.1 Connectivity
PSNs must ensure uninterrupted connectivity in case of disasters. In disaster situations, the ICT infrastructure may be damaged with fewer nodes available to support high volumes of traffic causing overloading. Therefore, it is essential to define the specific roles of each integrated technology, such as PS agencies (police, fire brigade, ambulance, etc.), legacy PSNs (PMR/LMR), commercial cellular networks and other, before, during, and after disasters. The techniques of Self-Organizing Networks (SONs) can be employed to provide uninterrupted service to first responders during disasters [19, 20].

D2D communications can be the first option with aerial eNodeBs are deployed in the affected area [21]. An efficient solution for vertical handoff between D2D communications, aerial eNodeB, Satellite Communication, and Wireless LAN must be designed to ensure uninterrupted connectivity. Instead of completely relying on these infrastructure-centric solutions for PS, we can consider opportunistic networks [22, 23] in disaster situations. Haggle [24] is a typical example, which is an asynchronous data-centric network architecture based on the publish-and-subscribe model.

Unmanned Aerial Vehicles (UAVs), such as drones, can also be used for PSNs in the cases when normal 5G networks does not provide their services, such as in the case of disaster [25].

11.4.1.2 Interoperability
Interoperability is the ability of first responders from different PPDRs to communicate and share information in real time. The interoperability is mainly due to proprietary solutions from local PPDRs and lack of standards for PS communications. The use of

commercial cellular networks in PS is emerging as one possible solution to solve inter-operability problems. However, sharing communication infrastructure is not enough to completely solve the interoperability problems. All IP-based solutions and common backhaul like NG 911 are needed to interoperate different PPDRs and communication technologies to share networks, data, and resources in an efficient way.

11.4.1.3 Resource Scarceness

Due to the increased number of PS organizations with their own network infrastructure, PSNs may undergo resource scarceness. In addition, voice traffic registered over commercial cellular networks in the areas affected by disasters increases during and after disaster [26]. In future PSNs, sharing resources with commercial cellular networks can result in resource scarceness in disaster situations. Simply allocating more radio resources to PSNs does not solve the issue of spectrum efficiency. New techniques for dynamically allocating radio resources like SDN must be adopted. Moreover, techniques like cognitive radio can further help to improve the spectrum efficiency of PS communications.

11.4.1.4 Security

PSNs must be secured from malicious attacks [27]. The existence of interconnectivity and interoperability of different agencies, communication technologies, co-workers, video surveillance, D2D networks, and WSNs results in a complex automated information system and introduces new security risks in PSNs. The end-to-end security of inter-agency and cross-border communication will be a great challenge in future PSNs.

Network sharing with commercial cellular networks will further tighten security requirements in PSNs. In shared networks, PS services must have their own mechanism for user authentication and authorization, which should be managed by PS service providers. For better security, the traffic from different PS agencies must also be separated from each other. A better solution for interoperability will facilitate the implementation and enforcement of end-to-end security. Several layers of protection must be provided to secure PSNs including application such as physical layer security [28], data encryption [29], and management-interface security.

11.4.1.5 Big Data

Traditionally, the data used by PS officials was static in nature, such as the police department's record and the court information system etc. However, this is not the case with OSNs today, where there is a need to monitor issue-based social media traffic across the universe of social media users. Another challenge in personnel security screening, where names are searched against law enforcement databases as well as against social media and open source information. This pushes the responsibilities of PS officials into the territory of "big data", where the volume, variety, and velocity of data outstrip the ability of the unaided analyst to deal with it.

On the other hand, the emergency centres, such as 911, do not support technology beyond caller ID and location-based information. So the NG911 centres must be able to receive and analyze a wide array of media-rich contents because many individuals simply report the incident on social media without contacting 911 directly, assuming that someone else has already done so.

11.4.2 Enabling Technologies

The interoperability of different agencies, communication technologies, legacy and broadband PMR networks, video surveillance, D2D networks, and WSNs results in a complex automated information system that needs to be managed in an efficient way in order to accomplish the aforementioned challenges. The following technologies can offer guaranteed and optimized spectrum, end-to-end security, uninterrupted connectivity, and resilience.

11.4.2.1 Software-Defined Networking

SDN will be one of the key enablers for this multi-technology and multi-agency complex informative system of future PSNs. A central controller, possibly residing in NG911 centres, will help to make the intelligent decisions for technology selection in time of disaster. This will also help to separate traffic from different agencies, with controlled interconnection interfaces between agencies. It will further increase the spectrum efficiency of the system by dynamically allocating radio resources among different technologies (like D2D and cellular) and agencies.

11.4.2.2 Cognitive Radio Networks

Cognitive Radio Networks (CRNs) will help to efficiently utilize radio resources among Primary Users (PUs) and Secondary Users (SUs). In an underlay CRN setting, the SUs will be allowed to share the spectrum with the PUs on the condition that the interference observed at the PU is below a predetermined threshold. A practical example for such networks would be an indoor femto-cell where the users will be allowed to deploy femto Base Stations (BSs) that will have the ability to share the spectrum with macrocell users. Such networks will prove to have an optimized spectrum utilization with uninterrupted connectivity [30]. Moreover, cooperative beamforming in Multiple-Input and Multiple-Output (MIMO) CRNs will help to incorporate physical layer security in the nodes and protect information from possible eavesdropping and/or jamming, especially in D2D cases.

11.4.2.3 Non-Orthogonal Multiple Access

Non-orthogonal Multiple Access (NOMA) will address the increasing demands of high spectral/energy efficiency, high connectivity, and low latency of future generations such as 5G and beyond. Relative to conventional Multiple Access (MA) techniques, which are based on orthogonal resources, inclusive of frequency division MA, time division MA, code division MA, and Orthogonal Frequency Division MA (OFDMA), the key idea in NOMA is to allocate non-orthogonal resources to serve multiple users simultaneously, yielding a high spectral efficiency while allowing some degree of interference at receivers [31, 32].

11.5 Conclusion

The chapter provided a very broad coverage of standardization efforts on PSNs. It spans well beyond the recent standards ratified by 3GPP as part of defining the 5th generation of cellular communications. This choice reflects the fact that other SDOs and their earlier standards have played a vital role in shaping the recent standardization of communications for PSNs in 5G. Many challenges still remain open. Some challenges arise from the lack of a single well-adopted standard that necessitates a co-existence of multiple standards in the foreseeable future and the requirement for connecting them in a way to offer the best performance as well as efficient utilization of scarce radio spectrum. The future revisions of the 5G standards would continue to look into these aspects.

References

1 M. Mezzavilla, M. Polese, A. Zanella, A. Dhananjay, S. Rangan, C. Kessler, T. S. Rappaport, and M. Zorzi. Public safety communications above 6 GHz: Challenges and opportunities. *IEEE Access*, 6:316–329, 2018. ISSN 2169-3536. doi: 10.1109/ACCESS.2017.2762471.

2 Muhammad Usman, Anteneh A Gebremariam, Fabrizio Granelli, and Dzmitry Kliazovich. Software-defined architecture for mobile cloud in device-to-device communication. In *Computer Aided Modelling and Design of Communication Links and Networks (CAMAD), 2015 IEEE 20th International Workshop on*, pages 75–79. IEEE, 2015.

3 Muhammad Usman, Anteneh A Gebremariam, Usman Raza, and Fabrizio Granelli. A software-defined device-to-device communication architecture for public safety applications in 5G networks. *IEEE Access*, 3:1649–1654, 2015.

4 F Craig Farrill. FirstNet Nationwide Network (FNN) proposal. *First Responders Network Authority, Presentation to the Board*, 2012.

5 Ramon Ferrus and Oriol Sallent. Extending the LTE√/LTE-A business case: Mission-and business-critical mobile broadband communications. *IEEE Vehicular Technology Magazine*, 9(3):47–55, 2014.

6 Stefan Parkvall, Anders Furuskar, and Erik Dahlman. Evolution of LTE toward IMT-advanced. *IEEE Communications Magazine*, 49(2), 2011.

7 Marija Buis. LTE-Advanced Release-10 Features Overview. *adare GmbH*, 2012.

8 Sassan Ahmadi. *LTE-Advanced: a practical systems approach to understanding 3GPP LTE releases 10 and 11 radio access technologies*. Academic Press, 2013.

9 Martin Sauter. *From GSM to LTE-advanced Pro and 5G: An introduction to mobile networks and mobile broadband*. John Wiley & Sons, 2017.

10 JS Roessler. Lte-advanced (3gpp rel. 12) technology introduction white paper. *München: Application Note-1MA252-Rohde & Schwarz International*, 2015.

11 Zeeshan Kaleem, Muhammad Yousaf, Aamir Qamar, Ayaz Ahmad, Trung Q Duong, Wan Choi, and Abbas Jamalipour. Public-safety LTE: Communication services, standardization status, and disaster-resilient architecture. *arXiv preprint arXiv:1809.09617*, 2018.

12 Christian Hoymann, David Astely, Magnus Stattin, Gustav Wikstrom, Jung-Fu Cheng, Andreas Hoglund, Mattias Frenne, Ricardo Blasco, Joerg Huschke, and Fredrik Gunnarsson. LTE release 14 outlook. *IEEE Communications Magazine*, 54(6):44–49, 2016.

13 Ramon Ferrus and Oriol Sallent. *Mobile broadband communications for public safety: The road ahead through LTE technology*. John Wiley & Sons, 2015.

14 Open Mobile Alliance. Push to talk over Cellular (PoC)-Architecture. *OMA-AD-PoC-V1 0-20060609-A*, 2006.

15 Mark H Zentner, Brian J Frommelt, George J Horwath, Paul F Mafera, and Daniel J Sawicki. Method to provide context-aware linkage between NG9-1-1 SMS and public safety incident, September 29 2015. US Patent 9,148,771.

16 Gianmarco Baldini, Stan Karanasios, David Allen, and Fabrizio Vergari. Survey of wireless communication technologies for public safety. *IEEE Communications Surveys & Tutorials*, 16(2):619–641, 2014.

17 Itsuma Tanaka and Takashi Koshimizu. Overview of GSMA VoLTE Profile. *NTT Docomo Technical Journal*, 13(4):45–51, 2012.

18 Daniel Câmara and Navid Nikaein. *Wireless Public Safety Networks Volume 1: Overview and Challenges*. Elsevier, 2015.

19 Kelly LeRoux and Jered B Carr. Prospects for centralizing services in an urban county: Evidence from eight self-organized networks of local public services. *Journal of Urban Affairs*, 32(4):449–470, 2010.

20 Venkatasantosha Bhargava Thondapu. Self-organizing networks prospects for group scheduling for disaster relief and public safety communications in cellular networks. Master's thesis, University of Missouri, Kansas City, Missouri, USA, 2014.

21 Javad Zeraatkar Moghaddam, Muhammad Usman, and Fabrizio Granelli. A device-to-device communication based disaster response network. *IEEE Transactions on Cognitive Communications and Networking*, 2018.

22 Zongqing Lu, Guohong Cao, and Thomas La Porta. Networking smartphones for disaster recovery. In *Pervasive Computing and Communications (PerCom), 2016 IEEE International Conference on*, pages 1–9. IEEE, 2016.

23 Luciana Pelusi, Andrea Passarella, and Marco Conti. Opportunistic networking: Data forwarding in disconnected mobile ad hoc networks. *IEEE communications Magazine*, 44(11), 2006.

24 James Scott, Jon Crowcroft, Pan Hui, and Christophe Diot. Haggle: A networking architecture designed around mobile users. In *WONS 2006: Third Annual Conference on Wireless On-demand Network Systems and Services*, pages 78–86, 2006.

25 Arvind Merwaday and Ismail Guvenc. UAV assisted heterogeneous networks for public safety communications. In *Wireless Communications and Networking Conference Workshops (WCNCW), 2015 IEEE*, pages 329–334. IEEE, 2015.

26 Anteneh A Gebremariam, Muhammad Usman, Riccardo Bassoli, and Fabrizio Granelli. SoftPSN: Software-defined resource slicing for low-latency reliable public safety networks. *Wireless Communications and Mobile Computing*, 2018, 2018.

27 Marius Portmann and Asad Amir Pirzada. Wireless mesh networks for public safety and crisis management applications. *IEEE Internet Computing*, 12(1), 2008.

28 J. M. Hamamreh, H. M. Furqan, and H. Arslan. Classifications and applications of physical layer security techniques for confidentiality: A comprehensive survey. *IEEE Communications Surveys Tutorials*, pages 1–1, 2018. ISSN 1553-877X. doi: 10.1109/COMST.2018.2878035.

29 Muhammad Rizwan Asghar, Ashish Gehani, Bruno Crispo, and Giovanni Russello. PIDGIN: Privacy-preserving interest and content sharing in opportunistic networks. In *Proceedings of the 9th ACM symposium on Information, computer and communications security*, pages 135–146. ACM, 2014.

30 Imran Shafique Ansari, Mohamed M. Abdallah, Mohamed-Slim Alouini, and Khalid A. Qaraqe. A performance study of two hop transmission in mixed underlay RF and FSO fading channels. In *Proceedings of the 2014 IEEE Wireless Communications and Networking Conference (WCNC' 2014)*. IEEE, 2014.

31 O. Kucur, G. K. Kurt, M. Z. Shakir, and I. S. Ansari. Nonorthogonal multiple access for 5G and beyond. *Hindawi Wireless Communications and Mobile Computing*, pages 1–2, 2018.

32 H. Lei, J. Zhang, K.-H. Park, P. Xu, I. S. Ansari, G. Pan, B. Alomair, and M.-S. Alouini. On secure NOMA systems with transmit antenna selection schemes. *IEEE Access*, 5:17450–17464, 2017.

12

Future Perspectives

Muhammad Ali Imran, Yusuf Abdulrahman Sambo and Qammer H. Abbasi

This book presents how 5G can support vertical industries to increase efficiency and open new productivity domains. Given that 5G is indeed a revolution and not merely an evolution of mobile communications, it implies that a complete redesign of mobile communication networks is on the cards. Unfortunately, most of the gains 5G is expected to deliver would be concentrated in urban and suburban areas, with rural areas often being neglected, especially in terms of mobile data connectivity. In this concluding chapter, a case is made for the provision of mobile data connectivity in rural areas by pointing out the benefits of rural deployment as well as the role of government in fast-tracking roll-out. Moreover, with changes coming to mobile networks sooner rather than later – 5G rollout is expected to start in 2020 – it is imperative to highlight future trends in wireless communications and how they will further the gains of, and complement, 5G. Hence, an evaluation of key technologies that are worth considering in the development of beyond 5G mobile communication systems is also presented in this chapter.

12.1 Enabling Rural Connectivity

Several technologies, including the use of aerial platforms, have been proposed over the years to expand rural connectivity; however, the goal of this section is not to focus on these technologies but to make a case for improving rural connectivity. Despite the increased deployment of mobile networks to almost ubiquitous levels, coverage still remains an issue as over 30% of the global population lacks access to cellular data connectivity. According to a June 2017 Ofcom report, outdoor mobile data coverage in the UK was only 63%, leaving substantial parts of rural areas without mobile data connectivity. The main challenge in providing coverage to these areas is the unfavourable cost-benefit scenario. Mobile network operators are incentivized by profit-making and the high costs of deploying and maintaining network infrastructure to cater for sparsely populated rural areas makes little business sense.

Although the United Nations has declared the Internet as "a basic human right and that no one should be denied access to the digital dividend that the Internet offers", it is important to look beyond providing data services to rural areas in individual cases and have a holistic view. Rural areas are indeed generally under-populated and can be

Enabling 5G Communication Systems to Support Vertical Industries, First Edition.
Edited by Muhammad Ali Imran, Yusuf Abdulrahman Sambo and Qammer H. Abbasi.
© 2019 John Wiley & Sons Ltd. Published 2019 by John Wiley & Sons Ltd.

considered to be unprofitable for mobile network operators to deploy their services. Nevertheless, the opportunities that 5G is expected to unlock in terms of enabling vertical industries should not be underestimated by the operators. Some of these industries, such as manufacturing, energy and agriculture, require large swathes of land that should ideally be located away from built-up and populated areas. Obviously, rural areas serve as the best choice for situating these industries, and are generally considered the home of energy plants and agriculture, in particular. This implies that 5G has the ability to increase the connection density of rural areas and, accordingly, profitability. This would result in a win-win-win scenario by increasing profitability for the mobile network operators, productivity for the vertical industries and stimulating the economy as well as ensuring energy and food security for the society.

Given the considerable levels of investment required to deploy 5G in rural areas, governments need to provide conducive environments by considering favourable regulatory and investment policies. Regulators need to adopt policies to enhance spectrum utilization by allowing operators to trade excess spectrum, grant sufficient spectrum and licences, and encourage infrastructure sharing to reduce OPEX costs of operators in rural areas. Governments should also consider tax incentives, ensure transparency, abolish multiple taxation, eliminate bureaucratic bottlenecks and provide a level playing ground for all operators.

12.2   Key Technologies for the Design of beyond 5G Networks

12.2.1   Blockchain

Although bitcoin is perhaps the most popular application of blockchain, there are numerous other applications of blockchain outside the realm of cryptocurrencies, which span across several sectors of the global economy. The basic idea of blockchain is having several distributed servers working jointly to sanction "transactions" between two parties in a verifiable and permanent way. Each record is called a block and multiple blocks are linked together using cryptography, such that each block contains a timestamp, transaction data and the hash of the previous block. In short, blockchain is a distributed ledger technology that enables open and trusted exchange of data without using a central trusted authority.

Blockchain has the advantages of eliminating single points of failure and improved security given the distributed authentication of transactions, which makes it harder for records to be altered by non-authorized third parties. Once data has been recorded in a block, it cannot be changed without altering all subsequent blocks and this requires the consent of the majority of servers in the network.

Although still at its infancy, especially in applications outside cryptocurrencies, blockchain has been receiving a fair share of interest from the wireless communications world. Perhaps the most talked about use case of blockchain in wireless communications is in the Internet of Things (IoT) domain. Blockchain appears to be the perfect enabler of IoT by facilitating peer-to-peer communication between IoT devices in a secure and cost-effective manner. For instance, machines in a connected-plant setting would be able to communicate and authenticate themselves using blockchain to enhance the production process with minimal human intervention and increased security. Other use cases of blockchain in mobile communications include radio access technology selection through autonomous blockchain-based smart contracts between devices and access points, subscriber authentication during roaming, smart cities and mobile money services.

12.2.2 Terahertz Communication

The carrier frequencies used for mobile communications have been increasing with each generation of mobile systems. Millimetre-wave (mmWave) communication over the 30 GHz to 300 GHz frequency range has been receiving a great deal of attention recently due to its huge bandwidth to support data hungry 5G and beyond applications. Similarly, researchers are focusing on optical communications to allow for even higher data rates and better physical security. Interestingly, there is an often-overlooked frequency band that sits just between mmWave frequency and optical frequency bands, the terahertz (THz) band. The THz band sits in the 0.1–10 THz frequency range and offers very large bandwidth in addition to many other attractive features including robustness, that would, theoretically, push capacity towards the Terabits per second range, an order of magnitude above mmWave systems.

The THz band bridges the gap between radio and optical frequency ranges by combining vital characteristics of both mmWave and optical frequency bands, or, simply put, the best of both worlds. When compared to mmWave, THz signals provide higher link directionality and higher security. Unlike optical frequencies, the THz band allows for non-line-of-sight propagation and has a better performance in unfavourable weather conditions. However, a major drawback of RF transmission at such high frequencies is increased attenuation, which limits the communication range; fortunately, the emerging network densification of 5G and beyond systems would result in having shorter communication link distances. This would facilitate the practical deployment of THz networks, with researchers even considering the use of THz communication in the uplink.

THz communication would allow for the creation of wireless nano-networks – a wireless network of nano-machines. Given the advances in nanotechnology and recent breakthroughs in advanced materials like grapheme and carbon nano-tubes, it is possible to design nano-scale antennas that could operate on the THz frequency band and be built on nano-machines. Nano-networks would facilitate Internet-of-Nano-Things for use cases such as environmental monitoring, lab-on-a-chip and industrial manufacturing.

12.2.3 LiFi

The frequency spectrum crunch is forcing migration towards higher frequencies in the radiofrequency spectrum. However, it is a known fact that the effect of path loss increases as the operating frequency becomes higher, which severely degrades received signals and network performance. Moreover, systems operating at these high frequencies must be designed to overcome signal propagation impairments such as blockages and shadowing by increasing line-of-sight probability through techniques such as beamforming and the use of small cells. These increase the complexity of wireless systems, energy requirements and backhaul infrastructure.

Given the numerous challenges of radiofrequency spectrum, researchers are currently focusing on light-based communication, termed LiFi. LiFi, which stands for Light Fidelity, is a form of visible light communication that promises high-speed wireless communication by using Light Emitting Diodes (LEDs) to encode data onto visible light by varying the intensity at which the LED flashes on and off. The shifting of the LED is done so quickly that is imperceptible to the human eye but detectable by a photo-detector. The unnoticeable changes in the quick switching of the LED bulbs are then converted to electrical signal by the photo-detector, resulting in theoretical speeds of over 10 GB/s from a single LED source.

LiFi has the advantage of utilizing current lighting infrastructure, including off-the-shelf LED bulbs; and unlike RF communication systems that require numerous circuits, antennas and complex receivers, LiFi uses simpler modulation techniques. The most obvious property of light is that it cannot travel through walls, which implies that LiFi signals can be contained within a physical space. This significantly reduces eavesdropping and increases privacy/security for the intended user(s). This property of light also makes it easier to avoid interference to/from neighbouring systems and can operate in areas where RF communication is restricted. However, LiFi is currently not thought to be suitable in the uplink, given that it involves transmitting visible light to the receiver.

12.2.4 Wireless Power Transfer and Energy Harvesting

There is a growing need for battery-free and ultra low-power devices, especially in domains like environmental monitoring and medical and industrial automation. This has paved the way for researchers to harness energy from ambient sources such as sunlight, mechanical, thermal or electromagnetic sources using a technique called Energy Harvesting (EH). EH allows for the design of low-power devices that can operate on demand or when they have accumulated enough charge from the neighbouring energy source, thereby eliminating the need for batteries and increasing the lifetime of these devices.

In order to push the gains of EH further, researchers are working on Wireless Transfer Power (WPT), which, as the name implies, deals with the transfer of electrical power over the wireless channel to be harvested by receivers. The EH receiver captures the energy transmitted by the WPT transmitter through the use of energy conversion circuits that consist of antenna(s), filters and rectifiers, and stores the harvested energy for later use. Low-power devices such as wearable body sensors, GPS and wireless earpods can harvest energy from nearby WPT sources. For instance, a mobile phone can wirelessly power a wearable body sensor and the sensor, in return, sends measurement data back to the phone via wireless communication protocols. WPT can also be used to power remote sensors that are distant from the WPT source for applications such as industrial control, automation and monitoring. It is worthy of note that the total transmitted power has to obey regulatory and safety standards.

Index

Enabling 5G Communication Systems to Support Vertical Industries, First Edition.
Edited by Muhammad Ali Imran, Yusuf Abdulrahman Sambo and Qammer H. Abbasi.
© 2019 John Wiley & Sons Ltd. Published 2019 by John Wiley & Sons Ltd.